THE MATH TUTOR'S FRIEND

systems of equations

sequences & series

determinants

exponents & logarithms

functions

matrices

BOOK 2

ALGEBRA I

by
Joe Cyr

Cyr, Joseph

The Math Tutor's Friend / *Book 2 – Algebra I* / by Joe Cyr

Third Edition ISBN: 978-1-967039-43-2 Joseph Cyr

Second Edition ISBN: 979-8-9879759-2-3 Joseph Cyr

First Edition ISBN: 9798464500280 Amazon Kindle, 2022

3 4 5 6 7 8 9

Subjects: Main category of book – *algebra; functions; matrices; determinants; exponents; logarithms; series; sequences* – high-school math

2.Other category - math tutor reference

Cover illustrations by Ramon E. Owen. Cover Design by Joe Cyr

Book design by Joe Cyr

The Math Tutor's Friend
Series

Books

1 - Pre-algebra
2 - Algebra I
3 - Algebra II
4 - Algebra III
5 - Probability & Statistics
6 - Geometry
7 - Trigonometry
8 - A Look at Calculus
9- Series Guide & "FLASHMATH" handout reference cards

Plus . . .
Books 1 through 8 contain test questions with answers explained

THE MATH TUTOR'S FRIEND

BOOK 2

ALGEBRA I

Sections

About the "Math Tutor's Friend" Series:

Teachers impart knowledge guided by lesson plans in a course syllabus. Math tutors, on the other hand, face varying topics at differing levels in successive sessions; for example, pre-algebra, advanced algebra, geometry, trigonometry, statistics, or even calculus! This series addresses the issue and is intended to provide the tutor with needed quick review of high-school math topics. The material is not presented rigorously nor does it abound with proofs and explanations that are available in textbooks - the tutor already knows this stuff - but a quick refresher before diving into a tutoring session is always a good idea.

Too many in high-school have reached the 11th and 12th grades lacking basic math skills such as memorizing the multiplication table (having to use a calculator to multiply 6 by 7, for example), handling signed numbers, and performing operations involving fractions - knowledge that should have been acquired by the fifth grade! I try to address this deficiency by devoting the first few tutoring sessions to these fundamentals and providing handouts of multiplication tables, etc.

Books 1 through 8 of this series include a section containing quiz questions (with answers explained) that the math tutor may find useful.

Book 9 is an overall index of topics. It contains a set of summary "FLASHMATH" pages for quick reference or for use as handouts.

This third edition corrects typos found in previous editions and expands discussion of some of the topics.

Happy Tutoring!

Joe Cyr

Sarasota, FL

•

THE
MATH TUTOR'S FRIEND
g(x)
asymptote
f(g(x))
f + g
[x]
It's easy !!
f(x)
function
f○g
f(x)/g(x)
hole
f•g
Mr. Math
MATH
FUNCTIONS

About Book 2: "Algebra I"

Section 1, *Math Functions.*

Emphasis -:

- Difference between functions and relations
- Function notation
- Vertical-line test
- Domain and range
- *Extrema* - local and global and inflection points
- Inflection point
- Graphing functions
- Types of functions
- Continuity
- Combining functions
- Rational functions
- Types of discontinuity

Section 2, *Linear Functions.*

Emphasis -:

- Concept of slope
- Forms of linear equations

Section 3, *Analysis of Math Functions.*

Emphasis -:

- Parent functions
- Symmetry
- Boundedness
- Asymptotes

Section 4, *Systems of Equations..*

Emphasis -:

- Methods for solving
- Distance formula - must memorize

Section 5, *Matrices & Determinants.*

Emphasis -:

- Notation
- Diagonal matrix
- Identity matrix
- Transpose
- Combining matrices
- Calculating values of determinants

Section 6, *Exponents & Logarithms.*

Emphasis -:

- Definition
- Law of Exponents
- Law of Logarithms
- Change of base

Section 7, *Sequences & Series..*

Emphasis -:

- Difference between series and sequences

• •

1. MATH FUNCTIONS

TABLE OF CONTENTS

(*continues . . .*)

TABLE OF CONTENTS (CONT'D)

•

1. MATH FUNCTIONS

Mathematical function

One may say in everyday language that a thing is a function of something else, like "Starting salary is a function of education," or "A planet's position is a function of time." We will speak here of mathematical functions, not vigorously, but enough to provide understanding at the high-school algebra level. See Figure 1-1. A *function* is like a "black box" that processes an input, x, and produces exactly one output, $f(x)$ [or y], as shown. We see immediately that f(x) [or y] is a dependent variable and x is an independent variable. That is, the value of y depends on the input value of x.

Figure 1-1

Function notation

A notation for functions is the *ordered pair* $(x, f(x))$, or commonly (x, y), because $y = f(x)$ (say "eff-of-eks"). So, given a function $f(x)$, if you state a value for x, then I can calculate y using the formula $f(x)$. More precisely, you must specify an allowable, value for x. For example, take the *square root function* $y = \sqrt{x}$ [equivalently, $f(x) = \sqrt{x}$]. You may not specify a negative number as x for the square root function because at our level of algebra, the square root of a negative number is undefined. The set of allowable values for x is the *domain* of the function. The set of y-values computed from the values of x in the domain is the *range* of the function (the range is sometimes call the *codomain*, but this is not always true). A function is uniquely represented by the set of all pairs (x, y) within the domain of the function. Each such ordered pair defines a point in the plane of the *Cartesian coordinate system* and falls on a curve called the *graph* of the function. Graphs of functions are commonly used in all business and scientific fields. Back to notation; "y" and "$f(x)$" mean the same thing and so are interchangeable; however "$f(x)$" is more flexible because it provides more information. One may say "$y = 2x + 3$; solve for y when $x = -1$". It is easier to say "$f(x) = 2x + 3$; find $f(-1)$" ["f of x equals $2x$ plus three; find f of negative one"]. You do exactly the same thing with either notation: replace x with -1 in the function, then multiply by the coefficient, 2, and then add 3, simplifying to get a final value of +1; that is, $f(-1) = 1$, i.e., $y = 1$.

Summary of function notation

The expression "$f(x)$" means "a formula, named f, has x as its input variable". It does not mean "multiply f by x" as in "ax" or "$a{\cdot}x$." Given $f(x)$, the *argument* x in the parentheses is the *independent variable*. As stated above, *y, or* $f(x)$, is the *dependent variable*, because its value depends upon the input value of x. Function names like $f(x)$, g(x), $h(x)$, $s(t)$, R(x), *etc*, allow more than one function in any single context.

Functions of two or more variables

Function notation is used for mathematical expressions of two or more variables, e.g., $f(w, x)$ or $f(x, y)$ or $g(w, x, y, z)$. Get the idea?

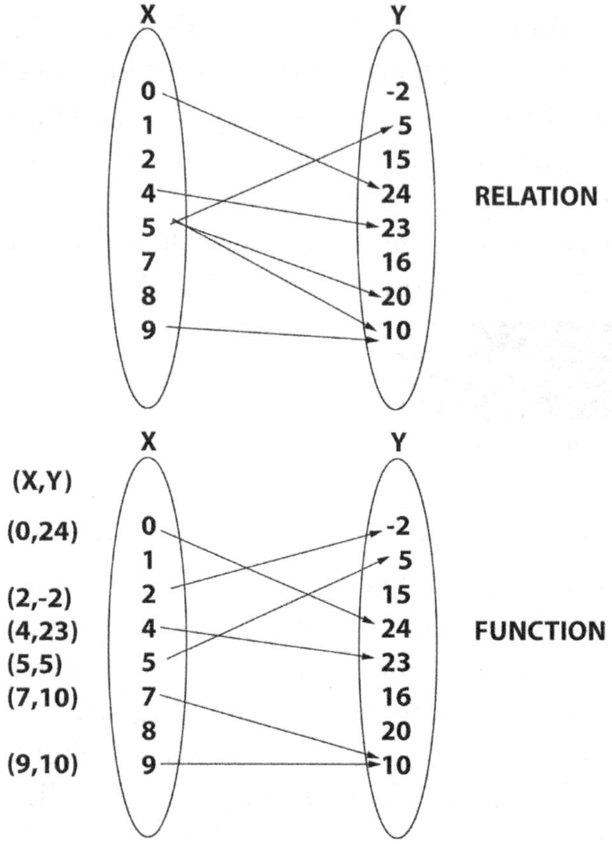

Figure 1-2

Sets, relations and functions

A *set* is a collection of objects (like a box of toys). Consider two sets of numbers: Set X and Set Y. See Figure 1-2. Set X consists of integers 0, 1, 2, 4, 5, 7, 8, 9 This may be written as X = {0, 1, 2, 4, 5, 7, 8, 9} and Y = {-2, 5, 15, 24, 23, 16, 20, 10}. X and Y are *unordered* sets; that is the order of the numbers in this set has no significance. We shall soon deal with ordered sets, where position is important. We will not get involved further with sets, recognizing that there are empty sets, subsets, finite and infinite sets, set building notation, union and intersection of sets, etc. Back to Figure 1-2; Sets X and Y each have eight elements. These are the *domain* and *codomain* of X and Y, respectively. There is a difference between a *function* and a *relation*. All functions are relations but all relations are not functions as we shall see. The relations between the elements of Set X and those of Set Y are shown as arrows in Figure 1-2. Integer 0 in X is related to 24 in Y in the RELATION as well as the FUNCTION. Note that 5 in Set X of the RELATION is related to 5, 10 and 20 in Set Y but that 5 in the FUNCTION is related only to 5 in Y. ***This is the distinction!*** A function relates a member of X to exactly one member of Y. The relationship of a function, $f(x)$ is defined by the *ordered pair* (x, y), where the first number is always x and the second number is always y. The ordered pairs for the FUNCTION are shown to the left in the figure. Some further observations: the members of Set X are the domain of the FUNCTION; the numbers -2, 5, 10, 23, and 24 in Set Y form the *range* while all of the members of Set Y are the codomain, that is, they are the possible values of Y for the FUNCTION. Note that the range differs from the codomain. Figure 1-3 shows the difference between a relation and a function.

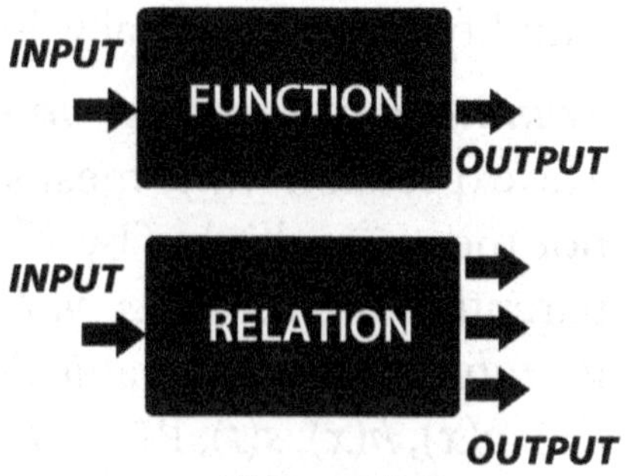

Figure 1-3

Graphing functions

One may graph a function using a graphing calculator or on one of the many Web sites that provide graphing services. My personal favorite is the Desmos Graphing Calculator at *https://www.desmos.com/calculator*. Most of the graphs in this book series were made on that site then copied using a snip tool. Graphing by hand can be done in two ways. If you just want to show the general shape, *extrema*, etc.. just plot a few point and then sketch it free-hand on a sheet of paper. If more accuracy is desired, as on a quiz, then the T-table, graph paper and a calculator may be used. See Figure 1-4.

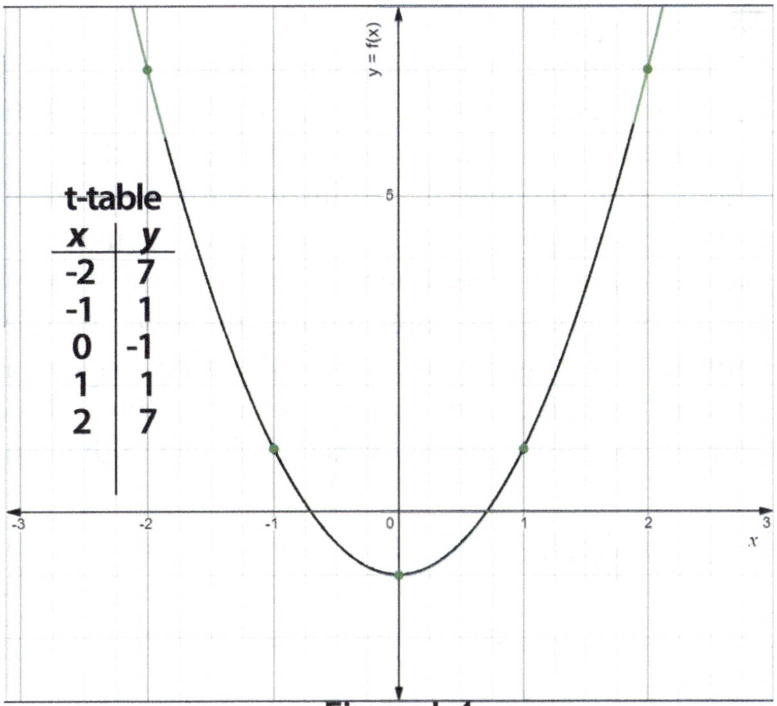

Figure 1-4

Vertical line test

This is a quick test to determine if a graph is indeed a function. Imagine a vertical line passing through the graph; if no vertical line can intersect at more than one point, then the graph is a function. This is the definition of a function: any specified *x* in the domain produces exactly one value of *y*.

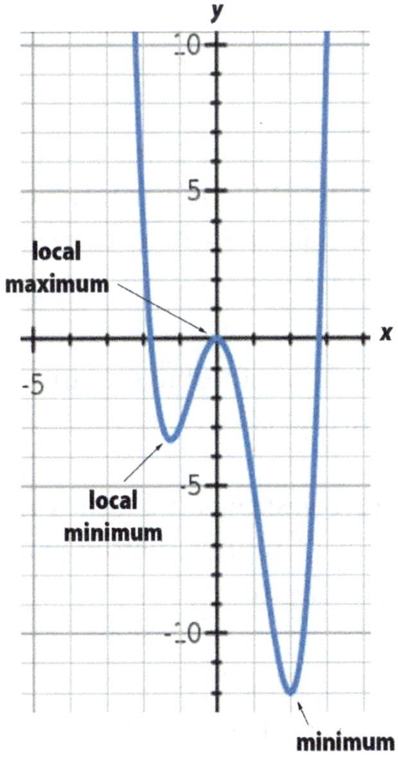

Figure 1-5

The Graphing calculator is a handy device that allows the user to enter several functions for superimposed display. Make sure that intended *modes* are set (dotted or continuous, function, degrees or radians, etc...).

Extrema

If a function is not a straight line (linear), then it will likely have a "highest" point, a "lowest" point and perhaps some "hills" and "valleys." These are *extrema* that may be *global* or *local*. Figure 1-5 shows three *extrema*: two local (a lower *extremum* and an upper *extremum)*. There is also a global *extremum* (global minimum) shown.

Inflection points

A *point of inflection* (*inflection point*) is a point at which the curve changes from *concave* to *convex*, or vice-versa. See Figure 1-6. Every defined point on a curve has an associated *slope* of the line tangent to the curve at that point (these can be found using differential calculus which is beyond the scope of this book). Now slopes may be positive, negative or zero. Moving from left to right on the graph, a slope decreasing to zero signifies a maximum at the point where the slope is zero; if we move along the graph and the the slope increases to zero (signifying a minimum), there must be an x-value where the slope changes from decreasing to increasing - this is the *inflection point* and defines where the function changes from concave to convex (or vice-versa) as shown in the figure.

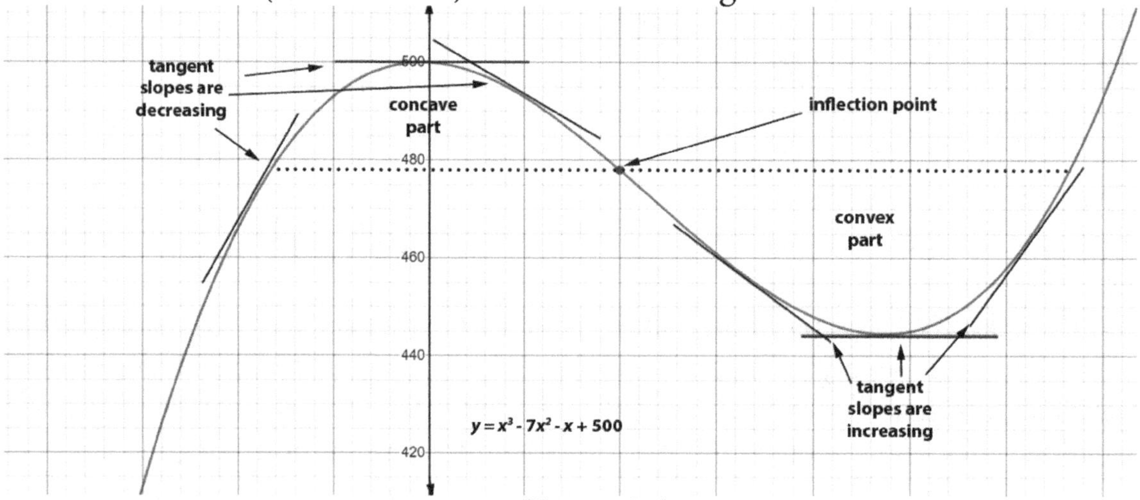

Figure 1-6

Types of functions

Linear function

A *linear function* is the equation of a straight line; one form of the linear function, the *slope-intercept form*, is $y = mx + b$ where $\boldsymbol{m}$ is the slope and $\boldsymbol{b}$ is the y-intercept. Figure 1-7, on the next page, shows the linear function $y = x + 2$.

Quadratic function

A quadratic function is a "U-shaped" curve with one *extremum* (global maximum or one global minimum). The *standard form of the quadratic* is

$$y = ax^2 + bx + c$$

where the coefficients a, b, and c are constants. Figure 1-7 shows the quadratic function $y = x^2 + 3x - 3$. Clearly, polynomial functions of higher degree exist, for example the function $y = x^3 - 5x$, also shown in the figure.

Trigonometric functions

An example of a trig function is the *sine* function $y = sin(\theta)$ where θ is an angle usually measured in either *degrees* or *radians*. Figure 1-7 shows the trig function $y = sin(x)$.

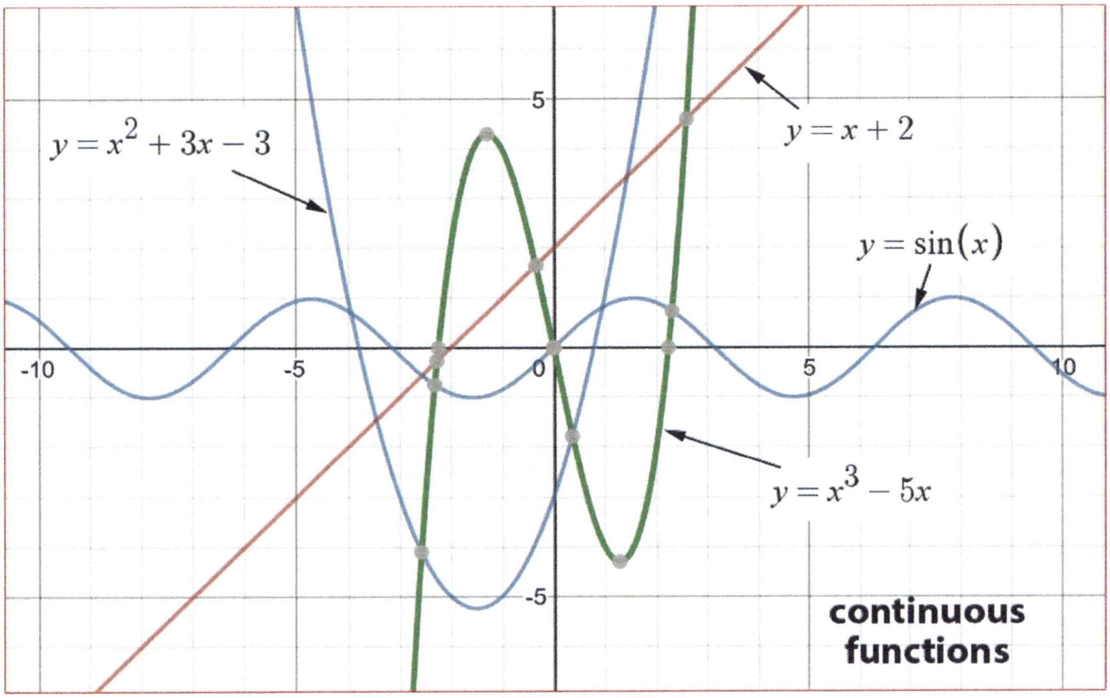

Figure 1-7

Other functions

Types of functions are virtually limitless: cubics, quartics, functions of two or more variables, etc. Figure 1-7 show a graph of the cubic function $y = x^3 - 5x$.

Continuous functions

Informally, if you can move a pencil along the function without having to lift it from the paper, then the function is *continuous*. Clearly, a function not continuous can be continuous withing a specified intervall of its domain.

Discontinuous functions

Rational functions (fractions with functions in both numerator and denominator) and other discontinuous functions have gaps in their domains. The gaps may be *holes* or *vertical asymptotes*. A *piecewise function* will partition the x-axis into intervals called *sub-domains* and define a separate function for each sub-domain. Asymptotes and a hole are shown in Figure 1-8.

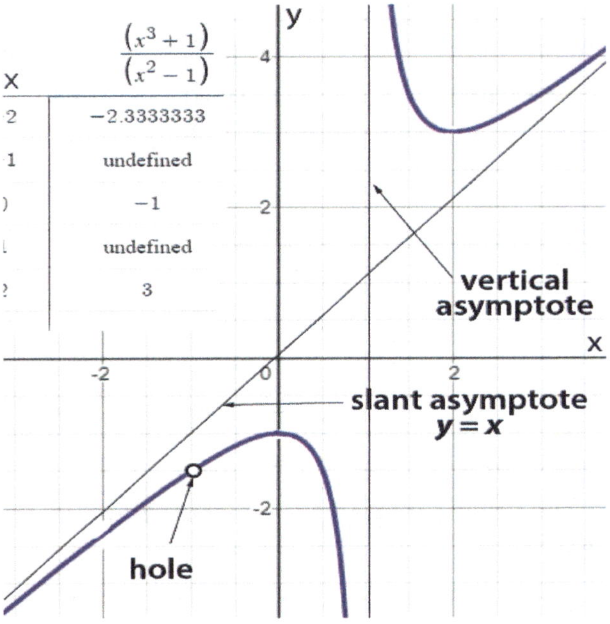

Figure 1-8

Combining functions

One may add, subtract, multiply, divide, or combine functions by composition.

Adding functions

There are several ways of showing the addition of functions. Consider two functions $f(x)$ and $g(x)$ to be added (or subtracted). Addition of functions may be shown as

$$f(x) + g(x) \text{ or } (f + g)(x) \text{ or } f + g;$$

i.e.,

$$f(x) + g(x) = (f + g)(x) = f + g.$$

Example: Given

$$f(x) = 3x^3 - 4x^2 + 2x - 1$$
$$g(x) = x^3 + x^2 + 1$$

find $(f - g)(x)$

When adding or subtracting two polynomials do the operation (plus or minus) on "like" terms; you may add or subtract only terms with variables of the same degree. For example, you may not combine $2x^3$ with $2x^2$, nor ax with x^2y. It is perfectly correct, however, to show the operation: $2x^3 + 2x^2$ and $x^2y + ax$. However, such sums can be simplified by factoring, a topic to be covered later. Returning to the example,

$$(f - g)(x) = f(x) - g(x)$$
$$= (3x^3 - 4x^2 + 2x - 1) - (x^3 + x^2 + 1)$$
$$= 2x^3 - 5x^2 + 2x - 2.$$

Multiplying functions

Multiplication of functions may be shown as $f(x)g(x)$ or $(fg)(x)$ or $f{\cdot}g$. That is $f(x)g(x) = (fg)(x) = f{\cdot}g$.

Example: Find $f{\cdot}g$ for the functions in the previous example.
The problem is to multiply $f(x) = 3x^3 - 4x^2 + 2x - 1$ by $g(x) = x^3 + x^2 + 1$

Multiplying two polynomials is straightforward but may entail a lot of work. One must multiply each term of the first polynomial by every term of the other polynomial and then collect *like terms*.

Example: $f{\cdot}g = f(x)g(x) = (3x^3 - 4x^2 + 2x - 1){\cdot}(x^3 + x^2 + 1)$

$$= 3x^3 \times x^3 + 3x^3 \times x^2 + 3x^3 \times 1 - 4x^2 \times x^3 - 4x^2 \times x^2 - 4x^2 \times 1 + 2x \times x^3 + 2x \times x^2 + 2x \times 1 - x^3 - x^2 - 1$$

$$= 3x^6 + 3x^5 + 3x^3 - 4x^5 - 4x^4 - 4x^2 + 2x^4 + 2x^3 + 2x - x^3 - x^2 - 1.$$

Collecting *like* terms yields

$3x^6 - x^5 - 2x^4 + 4x^3 - 5x^2 + 2x - 1$ in *standard form.*

Special case for binomials - FOIL

Clearly, multiplying polynomials is a lot of work with plenty of opportunities for errors, especially in signs. The principle is the same for multiplying two binomials, but using the mnemonic FOIL helps reduce the possibility of error. The acronym means "First, Outer, Inner, Last," explained in Figure 1-9 for multiplying the two binomials (a + b) and (c + d).

Example: Multiply $(x + 2)$ by $(x - 3)$.
Using FOIL we calculate $x^2 - 3x + 2x - 6$ which simplifies to $x^2 - x - 6$.

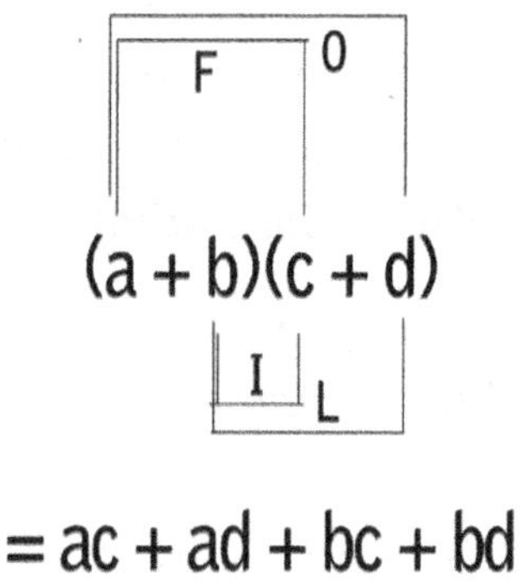

Figure 1-9

Dividing polynomials

One may divide a polynomial by a second polynomial. See the "long division" and synthetic division techniques discussed in Section 2, *Rational Functions*.

Function composition

Two functions $f(x)$ and $g(x)$ may be composed in two ways:

1. The composition of $f(x)$ and $g(x)$ (note the order) is, $(f\circ g)(x) = f(g(x))$
2. The composition of $g(x)$ and $f(x)$ is $(g\circ f)(x) = g(f(x))$

This means, simply, that $f(x)$ may be a function of $g(x)$ or vice-versa. A composition of two functions, say f(x) and g(x) is exactly like evaluating a function for a given value; i.e., "evaluate h(x) at $x = 4$," wherein we substitute "4" for x in h(x) and then evaluate. For the composite function f(g(x)), we substitute the entire function g(x) for each instance of x in f(x) and then simplify.

We need to note a couple of things here about function composition.

First, this is NOT multiplication. In other words, $f\cdot g \neq f\circ g$.
Second, the order we've listed the two functions is important $f\circ g \neq g\circ f$ in general, although there could be exceptions.

Examples:(Note that $f\circ g$ and $g\circ f$ yield different results)

Given $f(x) = x^2 + 2x - 1$ and $g(x) = 2x + 3$.

$f\circ g = f(2x + 3) = (2x + 3)^2 + 2(2x + 3) - 1$ which simplifies to
$f\circ g = 4x^2 + 12x + 9 + 4x + 6 - 1 = 4x^2 + 16x + 14$
$= 4(x^2 + 4x + 5)$
$g\circ f = 2(x^2 + 2x - 1) + 3$ which simplifies to
$g\circ f = g(x^2 + 2x - 1) = 2x^2 + 4x - 2 + 3 = 2x^2 + 4x + 1.$

Rational functions

Let $f(x)/g(x)$ be a *rational function*, which means that $f(x)$ is to be divided by $g(x)$ where $f(x)$ is of degree N with leading coefficient a_N and $g(x)$ is of degree D with leading coefficient a_D. A rational function where $g(x)$ contains the variable x is not continuous at values of x where g(x) = 0, may have one or more "holes" and will always have one or more *asymptotes*.

Holes

A *hole* exists in a rational function $f(x)/g(x)$ if there is a *common factor* in both numerator and denominator. For example, consider $(x - 1)/(x - 1)$; at $x = 1$, that ratio is the undefined 0/0, so the function has the *discontinuity* of a hole at $x = 1$.

Asymptotes

An *asymptote* is a line to which a function comes closer and closer but does not touch. We will discuss three types of asymptotes: *vertical asymptotes, horizontal asymptotes*, and *slant* (or *oblique*) *asymptote*. Vertical asymptotes are discontinuities in the domain. Figure 1-10 shows a vertical asymptote and a horizontal asymptote.

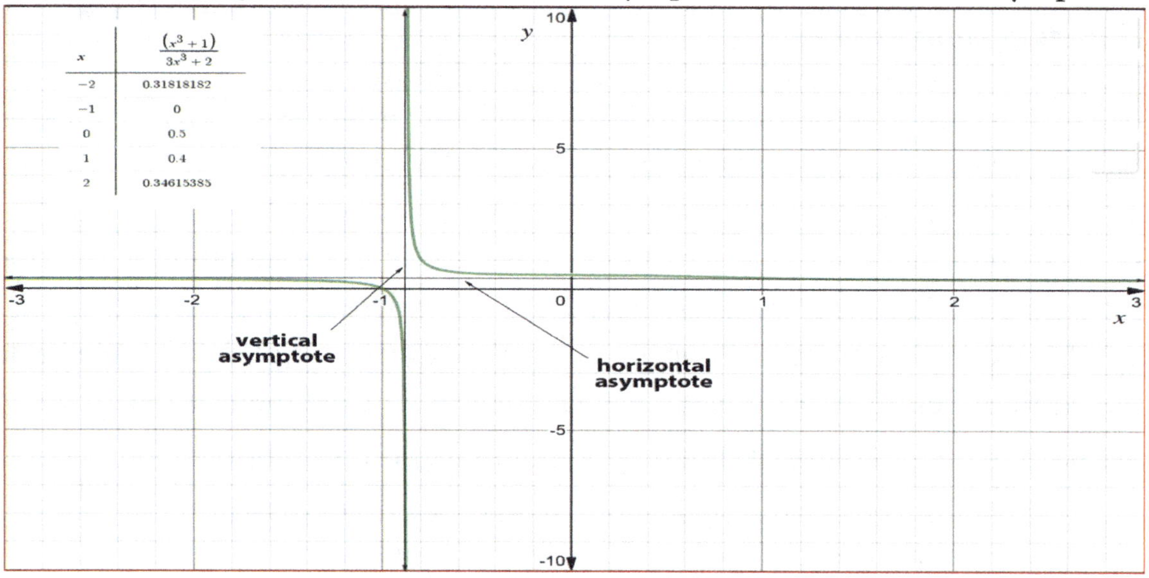

Figure 1-10

Horizontal and slant asymptotes

All rational functions include monomials and/or polynomials in both numerator and denominator. Let N be the *degree of the numerator* $f(x)$, that is, $N = deg[f(x)]$, and D the *degree of the denominator* $g(x)$, that is, $D = deg[g(x)]$.

Let's examine the four cases for $R(x) = f(x)/g(x)$: $N < D$; $N = D$; $N = D + 1$; $N > D + 1$.

Case 1: $N < D$. At large values of x (i.e., as $x \to \infty$), $R(x)$ tends to zero, therefore, there is a *horizontal asymptote* at $y = 0$ regardless of the values of a_N and a_D.

Case 2: $N = D$. At large values of x (i.e., as $x \to \infty$), $R(x)$ tends to the value of the ratio a_N / a_D, therefore there is a horizontal asymptote at $y = a_N / a_D$. See Figure 1-10.

Case 3: $N = D + 1$. There is a slant (or oblique) asymptote; use long division or synthetic division to divide the numerator by the denominator to obtain the linear equation of the asymptote ($mx + b$). See Figure 1-11.

Case 4: $N > D + 1$. There is neither a horizontal nor slant asymptote.

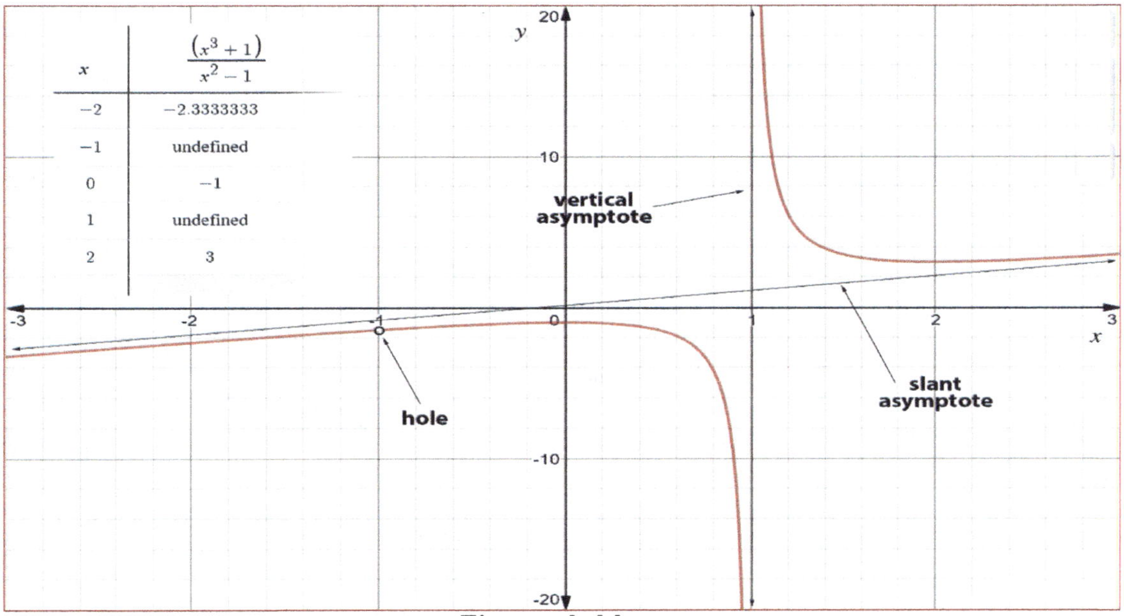

Figure 1-11

In summary, a rational function may have holes, vertical asymptotes, and a horizontal or slant (but not both) asymptote. The rational function $f(x)/g(x)$ may cross its horizontal or slant asymptote once. "Holes" represent *discontinuities* of the function and exist if both the numerator and denominator have identical factor(s). See Figures 1-12 and 1-13. A discontinuous function consists of *sub-domains* with a function defined within intervals on the x-axis). A function is defined for each sub-domain. See Figure 1-14.

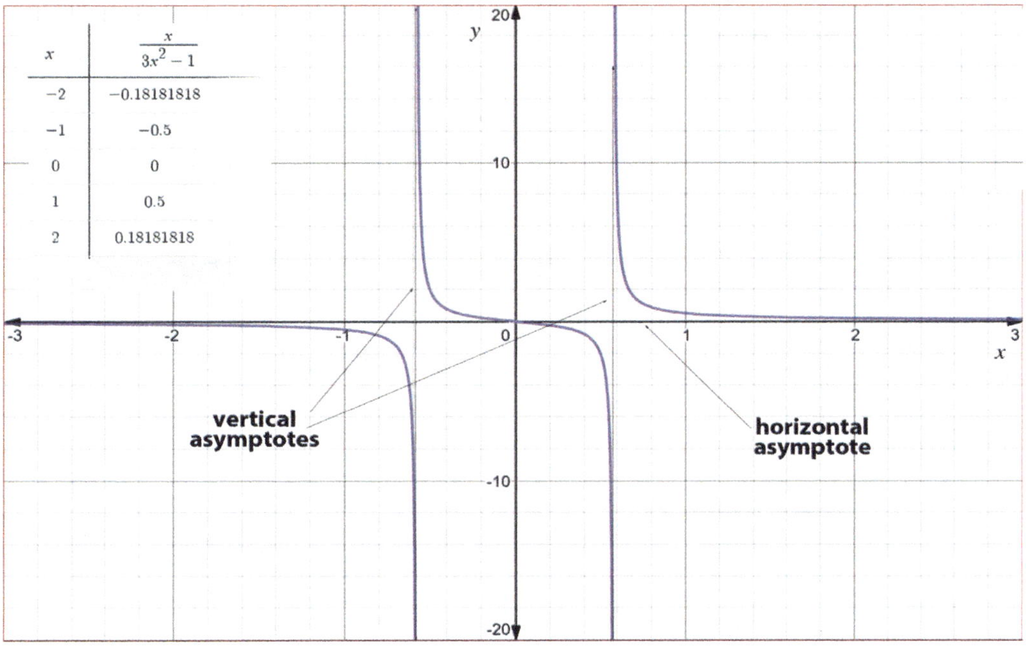

Figure 1-12

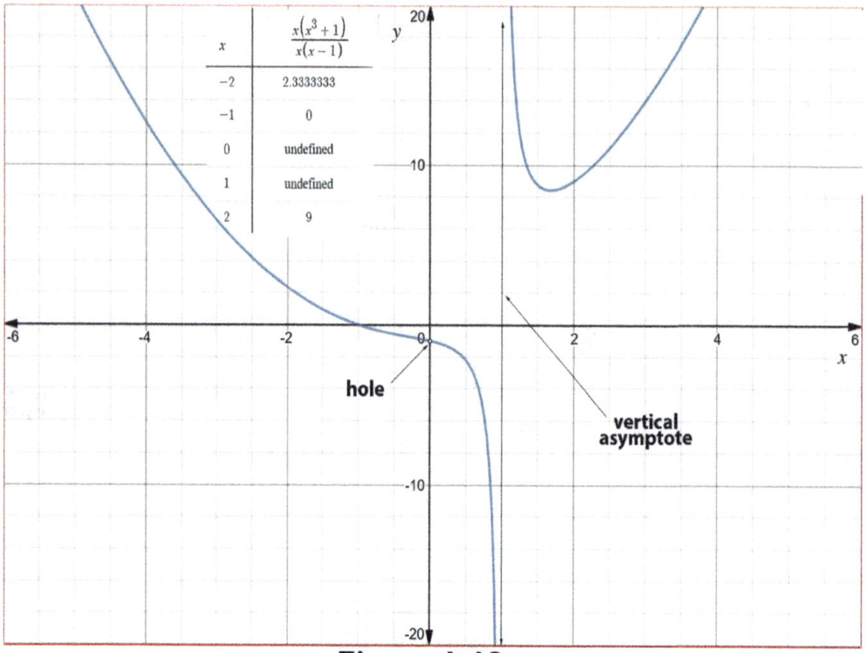

Figure 1-13

Piecewise functions (discussed at the end of this section)

An intuitive example of a *piecewise function* is the absolute value function $|x|$ which is a continuous function (why? It's defined throughout its domain).

$$|x| = x \; if \, x \geq 0$$

$$|x| = \text{-}x \; if \, x < 0$$

Just as with other functions, piecewise functions may be continuous over their defined intervals or may have holes and asymptotes. See Figure 1-14.

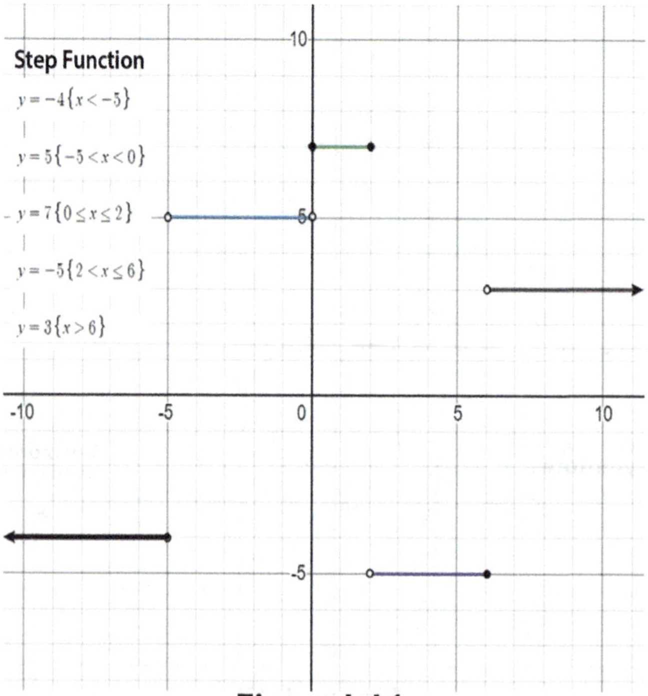

Figure 1-14

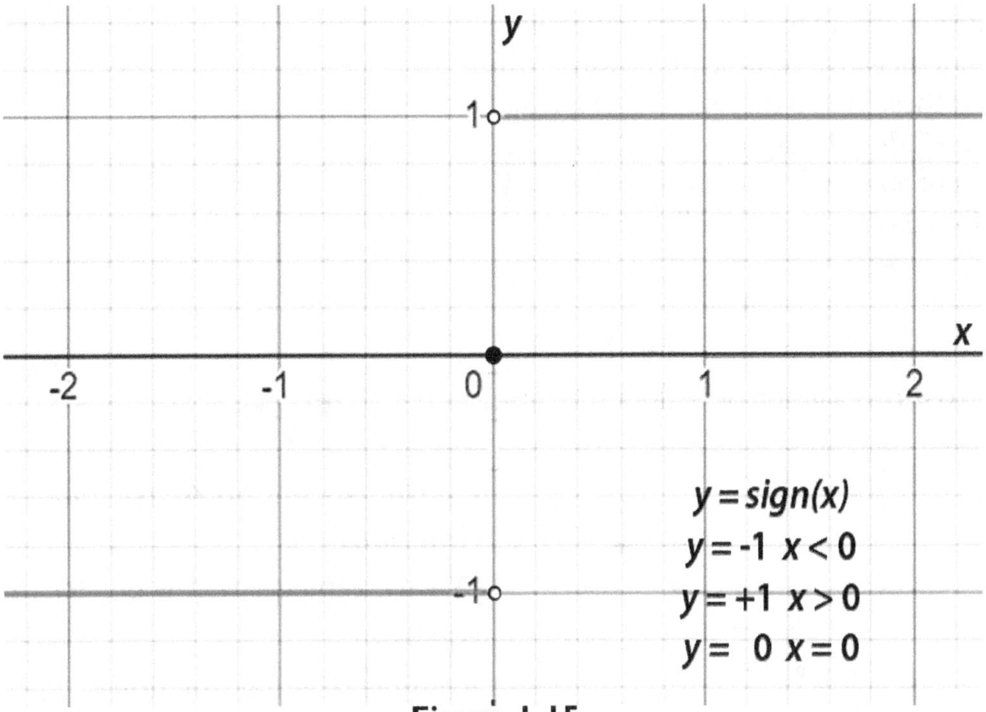

Figure 1-15

Step functions

A step (or staircase) function is a piecewise function having only finitely many pieces. Intervals are *pair-wise disjoint* (adjacent intervals have no point in common).
The constant function $f(x) = c$ where c is a constant is a step function. There is only one interval: $(-\infty, \infty)$. This is just a horizontal line at $y = c$.

The *signum* or *sign function** is -1 for negative numbers and +1 for positive numbers. It is the simplest non-constant step function. See Figure 1-15.

Rectangular function

also known as the *rectangle function*, *rect function*, *Pi function*, *gate function*, *unit pulse*, or the *normalized boxcar function*, is any function which is zero over the entire real line except for a single interval where it is equal to a constant, *c*. The transmitter pulse of a radar followed by a period awaiting target detection is an example of a rectangular function.

Staircase function

A function composed of a set of equally spaced jumps of equal length, such as the *ceiling function*, *ceil*(x), *floor function*, *floor*(x), or *nearest integer function*, *nint*(x). Let's look at the floor and ceiling functions.

Floor function (greatest integer function)

$\lfloor x \rfloor$ greatest integer that is less than or equal to x.

Example (Floor function): What it the floor of 3.14?
The floor is an integer that must be less than or equal to 3.14. There are an infinite num-

* Not to be confused with the ***sine*** function of trigonometry.

ber of integers less than 3.14 (remember, integers include negative numbers); however, the *greatest* of these is 3 so the *floor*(3.14) is 3 and does not include 4 (because 4 > 3.14). That is, 3 is the floor of any number in the interval [3, 4). This is plotted on a graph as follows:

•------------○

3 4

This is just one "step" of a step function. Figure 1-16 shows a graph of he floor function.

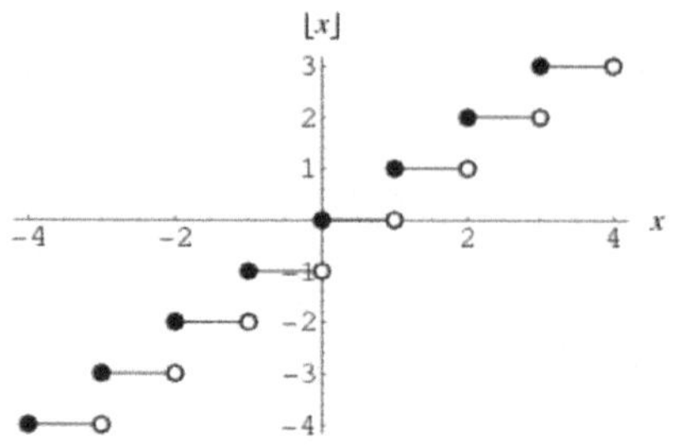

Figure 1-16

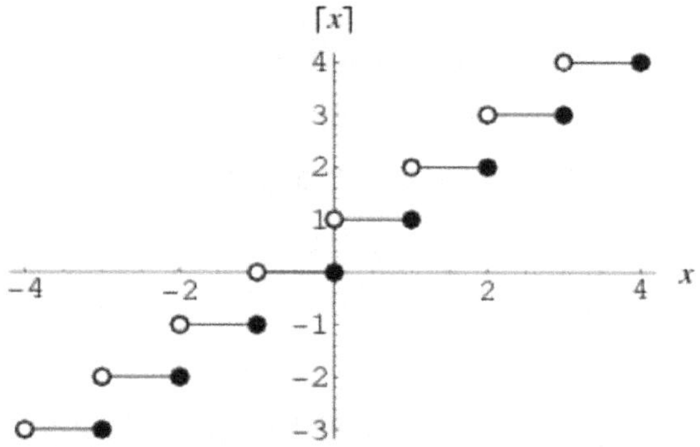

Figure 1-17

Ceiling Function (least integer function)

$\lceil x \rceil$: least integer that is greater than or equal to *x*.

<u>Example</u> (Ceiling function):
What is the ceiling of 3.14?

The ceiling is an integer that must be greater than or equal to 3.14.
As with the floor function, there are an infinite number of integers greater than 3.14, but the *least* of them is 4. So *ceil*(3.14) is 4 and does not include 3 (because 3 < 3.14). That is, 4 is in the interval (3, 4]. This is plotted on a graph as follows:

○------------•

3 4

This is one "step" of a step function. Figure 1-17 is a graph of the ceiling function. The ceiling function has been called the "gallows" because of the similarity in appearance of the symbol to that structure.

Floor and ceiling function symbology

The floor and ceiling function symbols are the square brackets ("[ ]") with the top or bottom serifs removed, respectively. The notation *floor*(*x*) and *ceil*(*x*) is also used.

Let m be an integer, then *floor*(x) = m if and only if $m \le x < m + 1$

<u>Example</u>: *floor*(3.14) = 3 because 3 < 3.14 < 4

ceil(x) = n if and only if $(n - 1) < x \le n$

<u>Example</u>: *ceil*(3.14) = 4 because (4 – 1) < 3.14 < 4

Fraction function

Just as *int*(3.14) = 3 [same as the *floor*(3.14)] is the integer part of 3.14; *frac*(3.14) = 0.14 Note: same as 3.14 minus *floor*(3.14), and same as 3.14 minus *int*(3.14)

Cautions in calculator use:

(1) When using a calculator do not use the "*FRAC*" feature for a number; it will not give the fractional part of a number but will convert the entire number to an improper fraction. For example on the TI-84, "3.14►*FRAC* = 157/50." Use the "*fPart(*" command instead (you may have to find it in the catalog of some calculators, e.g., the TI-84).

(2) Some calculators may give unexpected results for the *int* and *frac* functions. Test them with a number like (-3.14) to ensure that the results are what you want.

Many other types of functions exist, such as *congruence function, nearest integer function, box function, round function*, and so on. These are beyond the scope of this book.

Features of functions

Even and odd functions

A function may be *even* or *odd* depending upon its degree.

Even functions. A function, $f(x)$, is *even* if $f(-x) = f(x)$. *Even* functions of degree $2n$, $n = \mp 1, 2, 3, \ldots$, have a *global maximum* or *global minimum*. *Even* functions with positive leading coefficients originate and end at $+\infty$ and in addition to a global minimum may have local *extrema* (*local maximum* and *local minimum*); each local minimum [local maximum] must be followed by a local maximum [local minimum]. *Even* functions with negative leading coefficients originate and end at $-\infty$ and in addition to the global *extremum* (maximum) may have local *extrema*. Verify the above by plotting x^2 and $-x^2$.

Odd functions. A function, $f(x)$, is *odd* if $f(-x) = -f(x)$. *Odd* functions have a maximum at $+\infty$ and a minimum at $-\infty$, so must cross the x-axis at least once. *Odd* functions with negative leading coefficients start at $(-\infty, \infty)$ at $(\infty, -\infty)$; *Odd* functions with positive leading coefficients start at $(-\infty, -\infty)$ at (∞, ∞). Odd functions have no global *extremum* but may have local *extrema*.

An *odd* function may have local maxima and minima in pairs, that is, if originating from $-\infty$, if an *odd* function has a local maximum then it must also have a following local minimum to "change course" in order to proceed eventually to $+\infty$. Verify the above by plotting $(x^3 - 5x)$ and $-(x^3 - 5x)$.

Piecewise-defined functions

A *piecewise function* is one which is defined not by a single equation, but by two or more. Each equation is valid for some interval of the domain.

Example: The graph in Figure 1-18 is that of a piecewise-linear function because each of the three parts of the graph is a line. Piecewise-defined functions can also have discontinuities ("jumps").

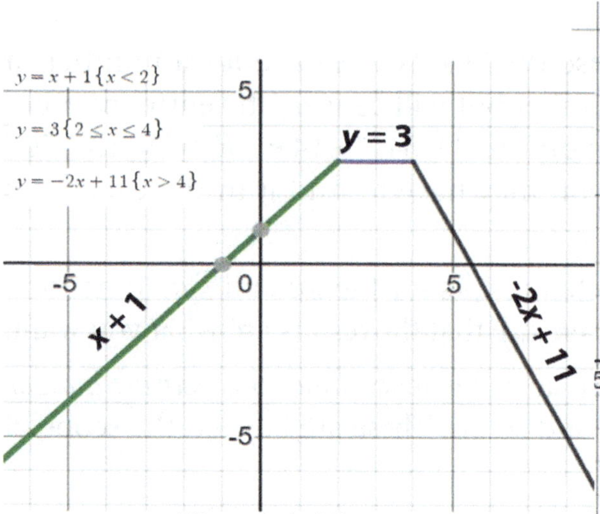

Figure 1-18

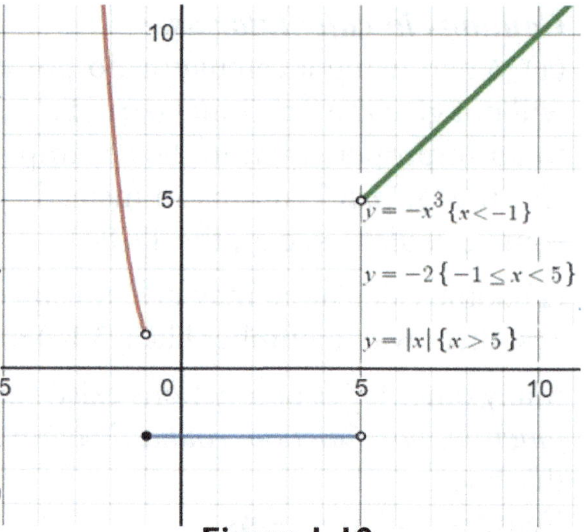

Figure 1-19

Example:

See Figure 1-19 above. Note that the small circles in the graph indicate that the endpoint of a curve is not included in the graph, and solid dots indicate endpoints that are included.

Example:

In the graph of Figure 1-20, negative values of x, as well as 0, are not included in the domain because the first function, $\log(x)$, is undefined for those values. The value $x = 10$ is not included in the domain because the second function is not defined for that value (it has a vertical asymptote there). Therefore the domain of this function is $(0, 10)$ and $(10, \infty)$.

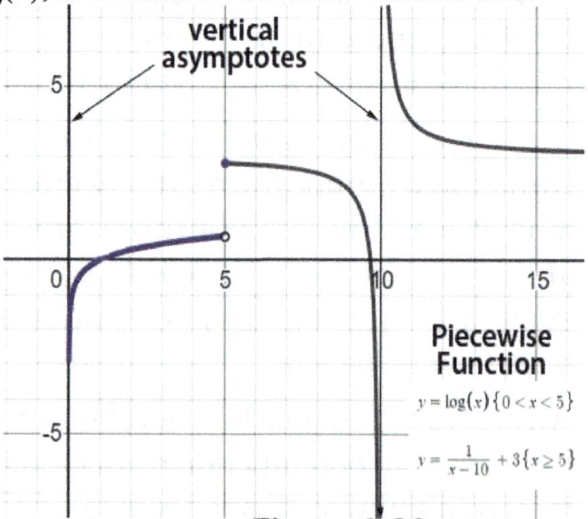

Figure 1-20

Parent functions

Many functions are simply *translations* or *dilations* (stretching or compressing) of basic functions called *parent functions.* Figure 1-21 on the next page lists the more common parent functions.

Translations of functions

Given a function $f(x)$, the function $f(x - h) + k$ translates $f(x)$ by h units to the right and k units up (if k and h are negative, then the translation is left and down, respectively). What about rotation? Well, visualize the function $y = f(x) = x^2$, which is "U-shaped." If you rotate $f(x)$ and then apply the vertical line test, you will find that you no longer have a function!

Inverse of a function

To find the *inverse of a function, i.e*, given $f(x)$ find $f^{-1}(x)$, just switch x and y then solve for y.

Example: Given $f(x) = (3x + 2)/5$ find $f^{-1}(x)$.

We have $y = (3x + 2)/5$, the equation of a line. To find $f^{-1}(x)$, switch x and y:

$f^{-1}(x)$: $x = (3y + 2)/5$; solving for y we get $y = (5x - 2)/3$ another line.

Therefore $f^{-1}(x) = (5x - 2)/3$.

Not all functions have inverses; more specifically, not all inverses of functions are themselves functions; that is, the inverse of a function is not necessarily a function. For example, $f(x) = x^2$ has as inverse $f^{-1}(x) = \pm\sqrt{x}$, which is not a function. So, how can you tell? Use the *horizontal line test*! If you can draw a horizontal line that intersects a function in more than one point, then that function does not have an inverse function. This is akin to the *vertical line test* to determine if a relation is a function.

PARENT FUNCTIONS

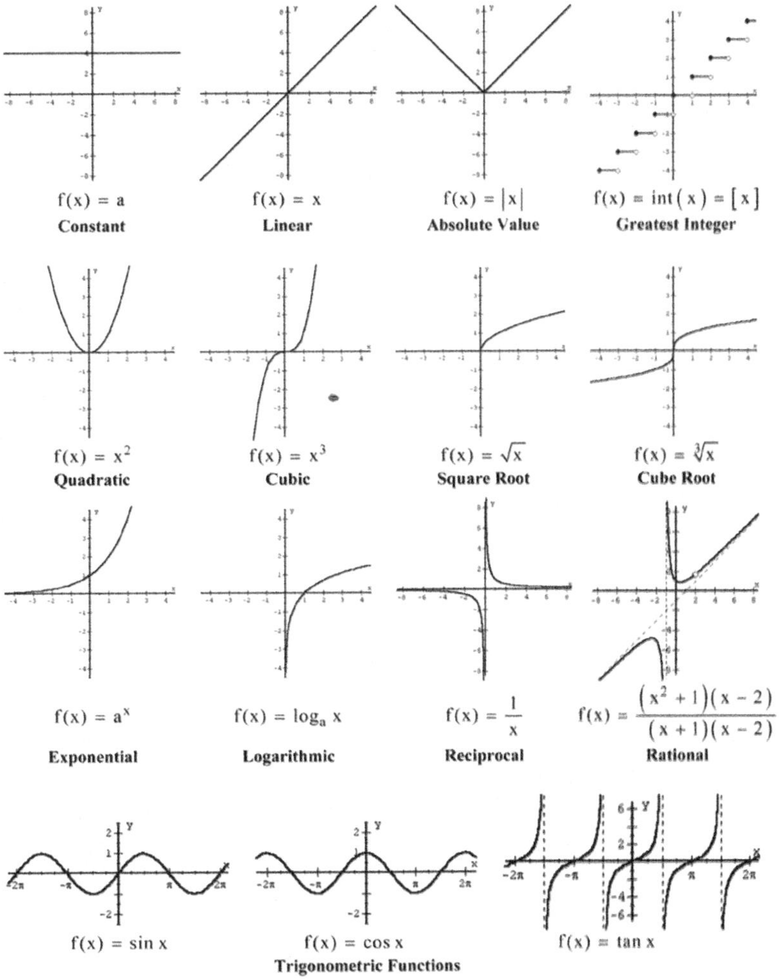

Figure 1-21

Reciprocal of a function

The *reciprocal* (*multiplicative inverse*) $f(x)^{-1}$ of a function $f(x)$ (or any number) is "1" divided by that function (or number), i.e., $1/f(x)$. In other words, the reciprocal of any fraction (recall that $f(x)$ is the same as $f(x)/1$) is that fraction "inverted." Thus the reciprocal of ¾ is ⁴⁄₃ and the reciprocal of $f(x)$ is $1/f(x)$.*

* $f^{-1}(x)$ and $f(x)^{-1}$ are NOT the same; $f^{-1}(x)$ is the INVERSE of $f(x)$, and $f(x)^{-1}$ is the RECIPROCAL of $f(x)$.

THE
MATH TUTOR'S FRIEND
y = mx + b
"rise over the run"
parallel lines
It's easy !!
y-intercept
Ax + By = C
LINEAR FUNCTIONS
slope
Mr. Math

2. LINEAR FUNCTIONS
TABLE OF CONTENTS

2. LINEAR FUNCTIONS

Coordinates of points on the plane

The *coordinates of a point* denoted by "(x, y)" are a pair of numbers that define the point's exact location on a two-dimensional plane with two axes: the horizontal *x-axis* and the vertical *y-axis*. The two numbers that form the coordinates of a point indicate its distance from the origin (0, 0) along each axis. The notation (x, y) is an *ordered pair*; that is, the first number, x, is always the distance of the point in the direction of the x-axis, and the second number, y, is its distance along the y-axis. Sometimes a letter "P" precedes the ordered pair to indicate a point, i.e., P(x, y).

From geometry we learn that any two points on a plane define a line; that is, one and only one line can be drawn through two points (x_1, y_1) and (x_2, y_2). The subscripts simply distinguish one variable from another. The graph shown below is that of a line through the two points (2, 2) and (6, 4), that is $(x_1, y_1) =$ (2, 2) and $(x_2, y_2) = (6, 4)$.

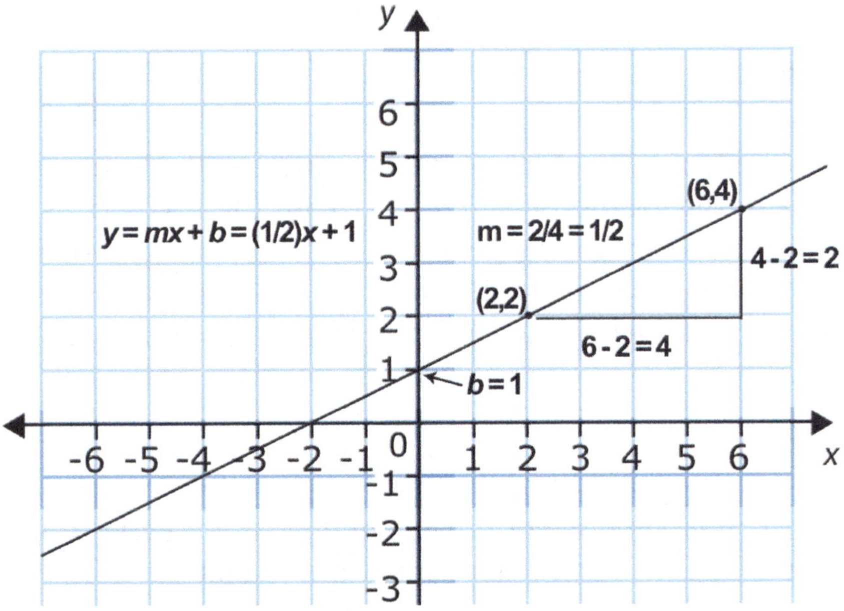

Graph 2-1

Introduction to the linear equation

A *linear equation*, or the *equation of a line*, is an equation with a single variable having an exponent of 1 (i.e., no exponent shown because $x^1 = x$).* For example,

$$y = mx + b$$

is a linear equation (*slope-intercept form*) with variable x and constants m and b; it is an equation of a line. The constant m is the *slope* of the line. Loosely

* A linear equation is an equation of degree 1, also called a first-degree equation.

speaking, the slope of a line is its direction or angle of rise or descent. To find the *slope*, you divide its rise by its run ("rise over the run"): the difference of the y-coordinates of any two points on a line (the "rise") divided by the difference of the x-coordinates of those same two points (the "run"). For Graph 2-1 above, the rise is the difference between y_2 and y_1 and the run is the difference between x_2 and x_1. Therefore, the slope of the line is $(4 - 2)/(6 - 2) = 2/4 = 1/2$. (Note: it makes no difference which point you designate as "1" or "2." The result will be the same, namely, the slope of the line). We have, then, $m = 1/2$. The graph clearly shows the *y-intercept* (where $x = 0$) and the *x-intercept* (where $y = 0$). Call the y-intercept "b." Thus, $b = 1$, as shown and $y = mx + b$ becomes $y = 1/2\, x + 1$. This is the *slope-intercept* form of the equation of a line, and is one that you should memorize.

Forms of the linear equation

There are many forms of linear equations of which seven are described in this section, namely, the *slope-intercept form* just discussed, the *point-slope form, two-point form, symmetric form, intercept form, general form*, and the *standard form*. Let's examine each of these using the same line shown in Graph 2-1. In addition to the seven forms listed above there are two special cases of lines: *horizontal* lines and *vertical* lines.

Slope–intercept form

$$y = mx + b \qquad (2\text{-}1)$$

As already discussed, m is the slope of the line and b is the y-intercept which is the y-coordinate of the location where line crosses the y axis. This can be seen by letting $x = 0$ which immediately gives $y = b$. Vertical lines have an undefined slope and cannot be represented by this form. From the previous discussion, the slope-intercept form of the line shown in the graph is $y = 1/2\, x + 1$. Compare this to the equation $y = mx + b$ to see that $m = 1/2$ and $b = 1$.

Point–slope form

$$y - y_1 = m(x - x_1) \qquad (2\text{-}2)$$

In Equation (2-2), m is the slope of the line and (x_1, y_1) are the coordinates of a given point on the line. The point-slope form shows that the difference between the y- coordinates of an arbitrary point (x, y) and the given point (x_1, y_1) is $y - y_1$; it is proportional to the difference between the x coordinates (that is, $x - x_1$). The *proportionality constant* is m, which is the slope of the line. The point-slope form in Graph 2-1 is

$$y - 2 = 1/2\,(x - 2).$$

Two-point form and symmetric form

Consider two points (x_1, y_1) and (x_2, y_2) on a line; the slope, m, of course, is the "rise over the run;" that is, the difference in the y values divided by the difference in the x values. Substitute these differences for m in Equation (2-2) produces the *two-point form*:

$$y - y_1 = \frac{y_2 - y_1}{x_2 - x_1}(x - x_1) \qquad (2\text{-}3)$$

Here (x_1, y_1) and (x_2, y_2) are two points on the line with $x_2 \neq x_1$. This is equivalent to the point-slope form in Equation (2-2) where the slope is explicitly given as $(y_2 - y_1)/(x_2 - x_1)$. For the line in Graph 2-1 we have

$$y - 2 = \frac{4-2}{6-2}(x - 2) = y - 2 = ½\,(x - 2).$$

Multiplying both sides of this equation by $(x_2 - x_1)$ yields a form of the line generally referred to as the *symmetric form*:

$$(x_2 - x_1)(y - y_1) = (y_2 - y_1)\,(x - x_1).$$

Substitute the numbers from the graph:

$$(6 - 2)(y - 2) = (4 - 2)\,(x - 2)$$
$$4(y - 2) = 2(x - 2)$$
$$2(y - 2) = (x - 2).$$

Intercept form

$$\frac{x}{a} + \frac{y}{b} = 1 \qquad (2\text{-}4)$$

where $a \neq 0$ and $b \neq 0$ (why?). The graph of the equation has x-intercept a, and y-intercept b. The intercept form can be converted to the standard form discussed below (Ax + By = C) by setting A = $1/a$, B = $1/b$ and C = 1. The example graph shows the y-intercept ($y = 1$, where $x = 0$), and the x-intercept is $x = -2$, where $y = 0$.

$$x/(-2) + y/1 = 1$$

or

$$y - x/2 = 1$$

which converts easily to the slope-intercept form.

Standard form

$$\mathrm{A}x + \mathrm{B}y = \mathrm{C} \qquad (2\text{-}5)$$

where A and B are not both equal to zero, A, B, and C are *coprime* integers (i.e., they have no common factors, also called *relatively prime*) and A is nonnegative (if A is zero, then B must be positive). Let's begin with the slope-intercept form and derive the standard form of Equation (2-5):

$y = mx + b$	[Slope-intercept form]
$-mx + y = b$	[Move mx to the left side of the equation]

Comparing the coefficients of the preceding equation to the coefficients of Equation (2-5), we see that A = $-m$; B = 1; and C = b. The standard form, then,

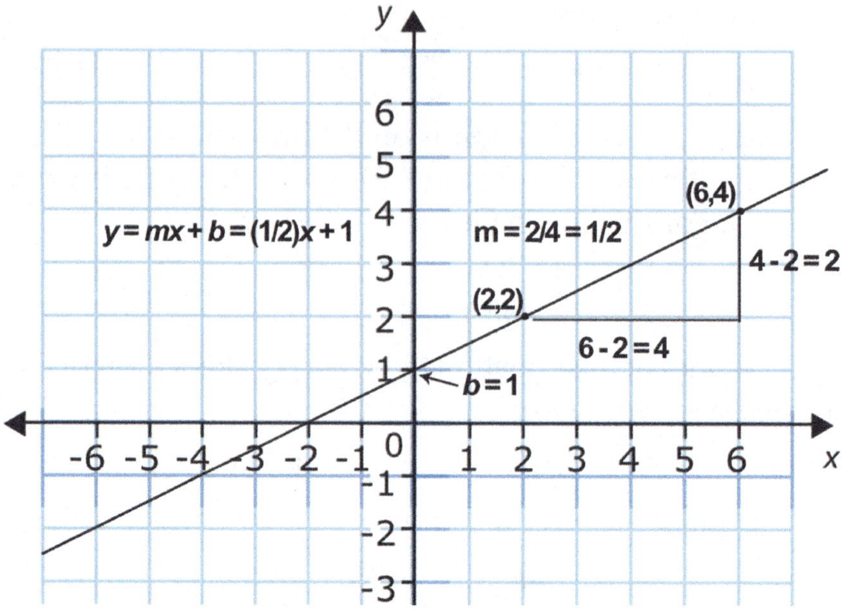

Graph 2-1

for the example line is

$$-\tfrac{1}{2}x + y = 1$$

General form of the linear equation

$$Ax + By + C = 0 \qquad (2\text{-}6)$$

where A and B are not both equal to zero (A is usually but not required to be non-negative). The graph of the equation is a straight line, and every straight line can be represented by an equation in the above form. If A is not zero, then the x-intercept, that is, the x-coordinate of the point where the graph crosses the x-axis (where, $y = 0$), is $-C/A$. If B is not zero, then the *y-intercept*, that is the y-coordinate of the point where the graph crosses the y-axis (where $x = 0$), is $-C/B$, and the *slope* of the line is $-A/B$. The slope-intercept form of the example line is $y = \tfrac{1}{2}x + 1$, as already determined above. To put this in the general form we simply put everything to the left of the "=" sign: $-\tfrac{1}{2}x + y - 1 = 0$; comparing this with the general form we have $A = -\tfrac{1}{2}$; $B = 1$; and $C = -1$.

Special cases (horizontal and vertical lines)

The *horizontal* line is a special case of the slope-intercept form with slope, $m = 0$:

$$y = mx + b = (0) \times (x) + b = b.$$

Therefore, $y = b$ is the equation of the horizontal line which is parallel to the x-axis and intercepts

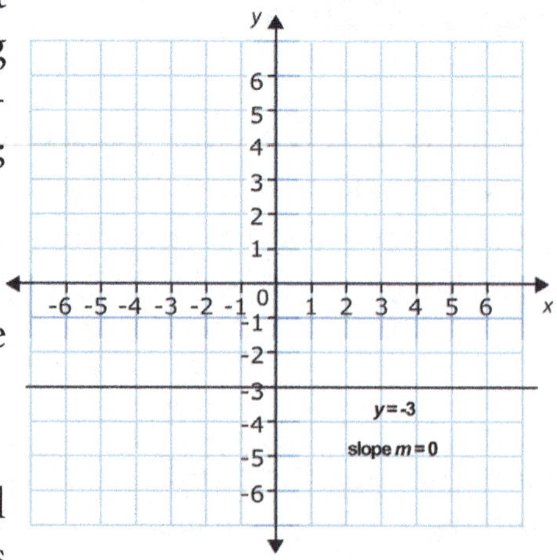

Graph 2-2

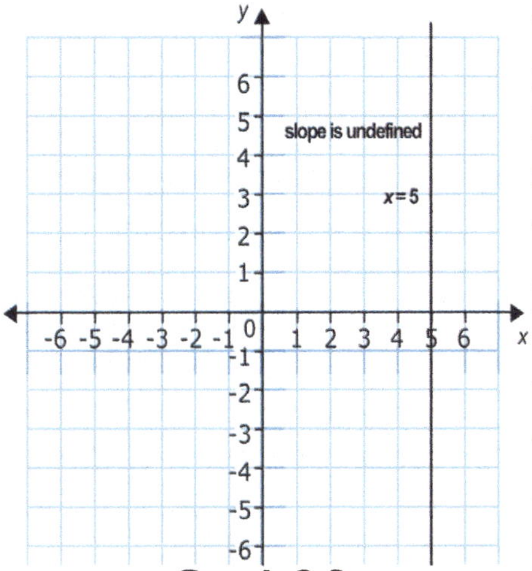

Graph 2-3

the y-axis at b. There is no x-intercept, unless $b = 0$, in which case the graph of the line is the x-axis. The horizontal line is also a special case of the standard form with $A = 0$ and $B = 1$.

The *vertical* line is given by $x = a$, a vertical line passing through x-axis at a. Thus $x = 5$ is a vertical line that passes through $x = 5$; if $x = 0$ the vertical line is the y-axis. This is a special case of the standard form ($Ax + By = C$) where $A = 1$ and $B = 0$. Graph 2-3 shows a vertical line with x-intercept equal to C. The slope is undefined. There is no y-intercept, unless $a = 0$, in which case the graph of the line is the entire y-axis.

Parallel lines

If two lines have identical slopes and different y-intercepts, then the lines are *parallel*. It's as simple as that! See Graph 2-4 below. The two lines shown have slope ½ and y-intercepts 1 and -4, respectively. The lines, therefore, are parallel.

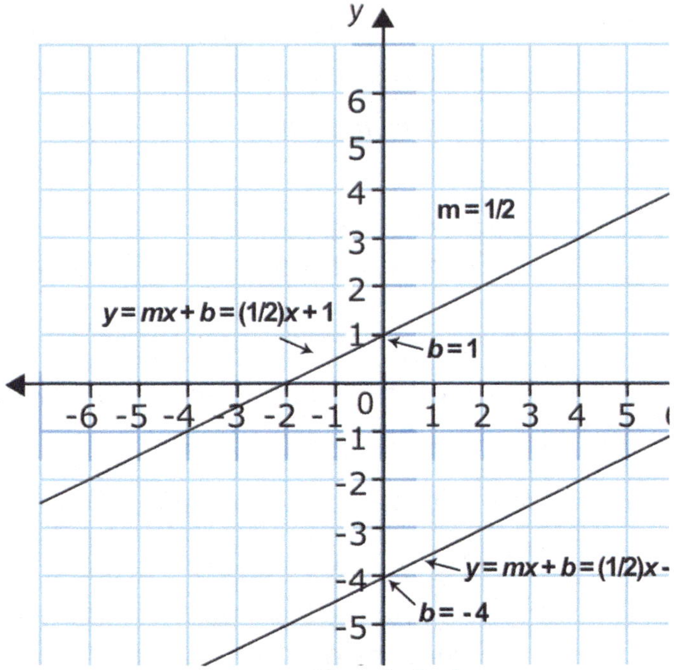

Graph 2-4

Perpendicular lines

A line *perpendicular* to (*normal to*) another line with slope m will have a slope $-1/m$. In words, a line whose slope is the *negative reciprocal* of the slope of another line is perpendicular to that line.

Distance of a point from a line

Let's do the easy cases first:

(1) Distance of a point $P(x_0, y_0)$ to horizontal line $y = b$. The y-value is a constant, b, so for any value of x, the distance to the line is $|y_0 - b|$. Because distance is always a positive number, the bars "| |" indicate *absolute value* which is always positive.

(2) Distance of a point $P(x_0, y_0)$ to a vertical line $x = k$. The x-value is a constant (k), so for any value of y, the distance to the line is $|x_0 - k|$.

(3) Distance of a point $P(x_0, y_0)$ to a line which is neither vertical or horizontal. The equation of a line is given by $y = mx + b$. The equation of a normal line to that line which passes through the point P is

$$y_0 = (-1/m)x_0 + b_0 \qquad (2\text{-}7)$$

The coordinates of the point $P(x_0, y_0)$ are known, so we must solve for the y-intercept of the normal line; therefore, solving for b_0:

$$b_0 = y_0 + x_0/m.$$

Substituting this into the equation of the normal line, Equation (2-7):

$$y = (-1/m)x + y_0 + x_0/m = (x_0 - x)/m + y_0.$$

The point at which these two lines intersect is the closest point on the original line to the point P, so we equate the two lines [they intersect at the point (x, y)]:

$$y = mx + b$$

and

$$y = (x_0 - x)/m + y_0.$$

Equating the two right-hand expressions and solving for x (do the work as an exercise) we get

$$x = [x_0 + m(y_0 - b)]/(m^2 + 1).$$

Substituting this in the equation of the original line,$y = mx + b$, yields

$$y = \{m[x_0 + m(y_0 - b)]/(m^2 + 1)\} + b$$

The formula for the distance, D, between two points (x, y) and $(x_{0,} \cdot y_0)$ is

$$D = \sqrt{(x_0 - x)^2 + (y_0 - y)^2} \qquad (2\text{-}8)$$

Now, substituting and simplifying the expressions for x and y into the distance formula [Equation (2-8)] involves considerable algebraic work, so we provide only the result for the distance of a point $P(x_0, y_0)$ to the line $y = mx + b$:

$$D = |b + mx_0 - y_0| / \sqrt{1 + m^2}\,.$$

Linear factors

One may factor functions such as quadratics (e.g., $ax^2 + bx + c$), cubics (e.g., $ax^3 + bx^2 + cx + d$), and higher-order functions into *linear factors* such as $(3x - 4)$, $(x + 2)$, or simply x. Each of these is a linear expression like those we have dealt with already in this section. The product of the linear factors of a function allows one to identify the *zeroes* (i.e., the x-intercepts) of the function, that is, the values of x where $y = 0$ (i.e., where the function crosses the x-axis). For example, the linear factors of the function

$$3x^3 + 2x^2 - 8x$$

are

$$x(3x - 4)(x + 2)$$

and so the three zeroes of the function are at $x = 0$, $x = 4/3$ and $x = -2$.

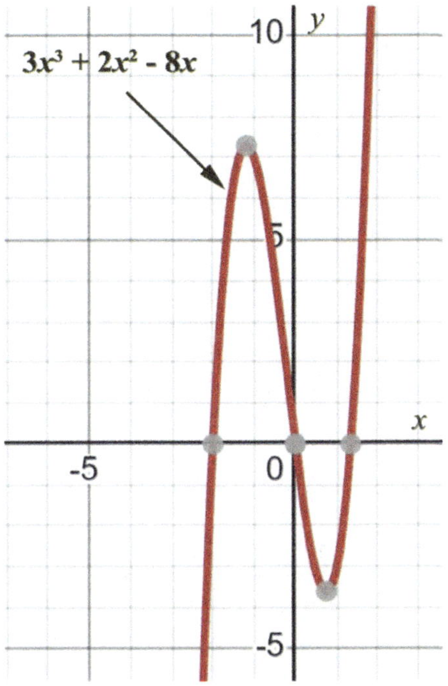

●

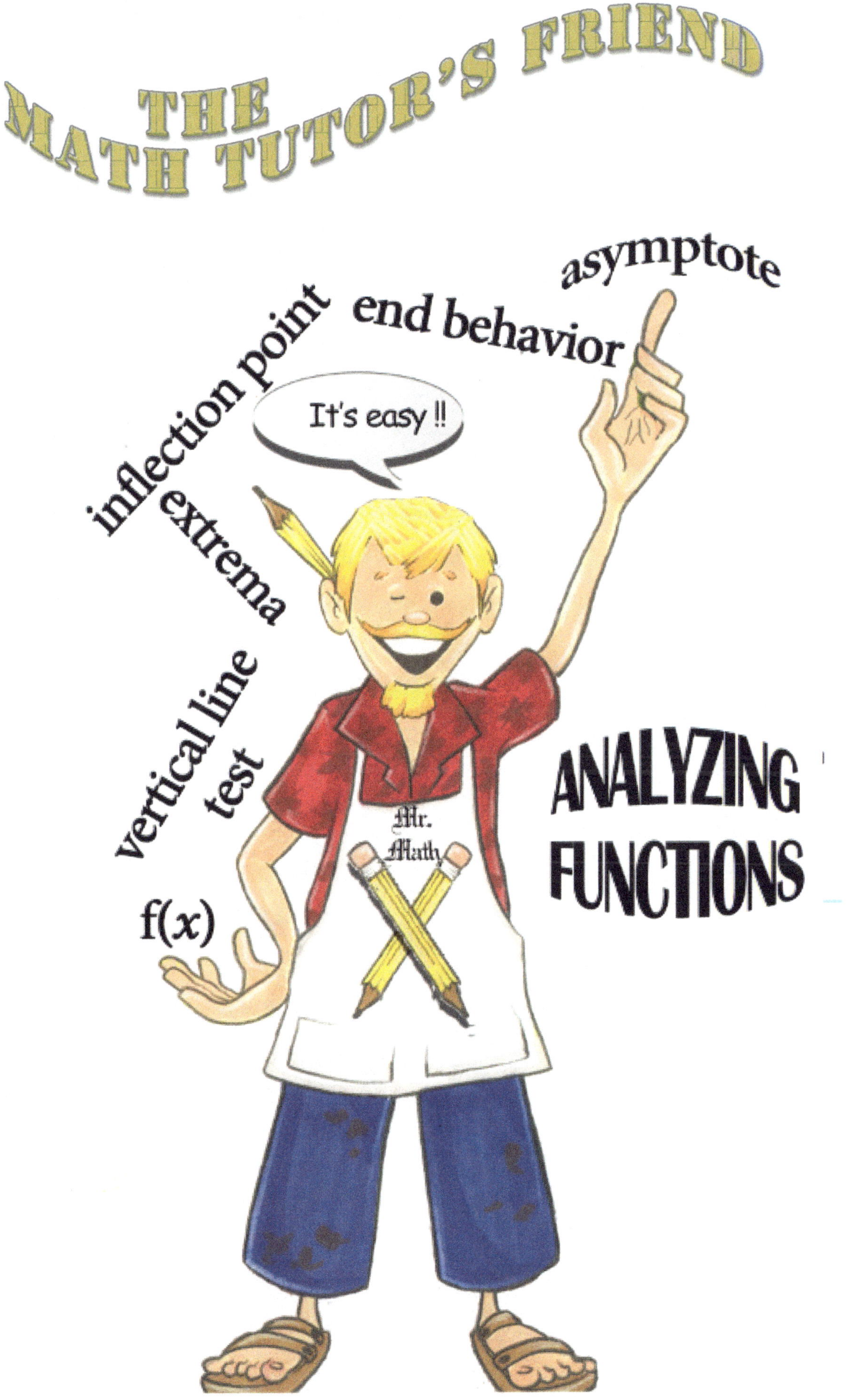
THE
MATH TUTOR'S FRIEND
asymptote
end behavior
inflection point
It's easy !!
extrema
vertical line
test
f(x)
Mr.
Math
ANALYZING
FUNCTIONS

3. ANALYZING MATH FUNCTIONS
TABLE OF CONTENTS

•

3. ANALYZING MATH FUNCTIONS

Graphing relations and functions

Graph 3-1 is a grid called the *Cartesian Coordinate system*. It provides a way to display the points of a relation as a graph. The Cartesian Coordinate system has a horizontal x-axis (*abscissa*) and a vertical y-axis (*ordinate*) that intersect each other at the *origin*. The axes are marked off in some convenient scale so that the (x, y) points of interest will appear on the observable part of the plane which is divided into four *quadrants* marked counterclockwise with Roman numerals as shown. The signs of x and y are organized as follows: (+, +) in Quadrant I; (–, +) in Quadrant II; (–, –) in Quadrant III and (+, –) in Quadrant IV. This is especially useful in the study of trigonometry.

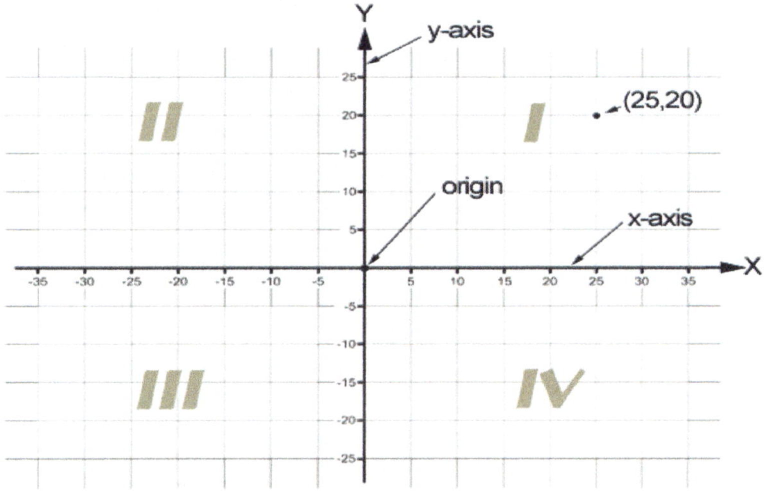

Graph 3-1

Setting up a coordinate system

Given the coordinates of points in (x, y) format, sketch the x and y axes and label* them. Next, determine the appropriate scales for the axes. The same scale is generally used for both axes; that is, each square or tick mark has the same value for both x and y. With the axes labeled and the scale set we are ready to plot the points of the function to be displayed. Graph 3-1 displays the point (25, 20).

Graphing calculators

Figure 3-1 on the next page shows two popular (*ca* 2020) calculators by Casio and Texas Instruments. Enter the function at the prompt " y= " and then press the "Draw" button to view the graph and x-y table for the displayed function.

* We deal only with x and y axes in this book However, it is important to label the axes of a graph because many types of axes are possible, e.g., t (time), $M (millions of dollars), d (distance), etc.

Figure 3-1

Graph of a function

Graph 3-2 below shows the function

$$f(x) = (x - 1)^2 - 1. \qquad (3\text{-}1)$$

Note that this is a function because each allowable value of x (the domain) is associated with exactly one value of y in the *range*. Note also that it is perfectly OK to have the same value of y for different values of x. The domain and range may be any real number in the range $(-\infty, \infty)$. This means that the domain (and range) can contain irrational numbers like $\sqrt{2}$ and transcendental numbers like π and e.

Now, let's look at another function that you may encounter in your study of algebra:

$$y = x^3 - 1 \qquad (3\text{-}2)$$

We know that Equation (3-2) is a function because for any specified value of x there will be exactly one value of y. See Figure 3-3. Now let's see how we can analyze that function. Visualize the vertical line test to confirm that the graph indeed is that of a function.

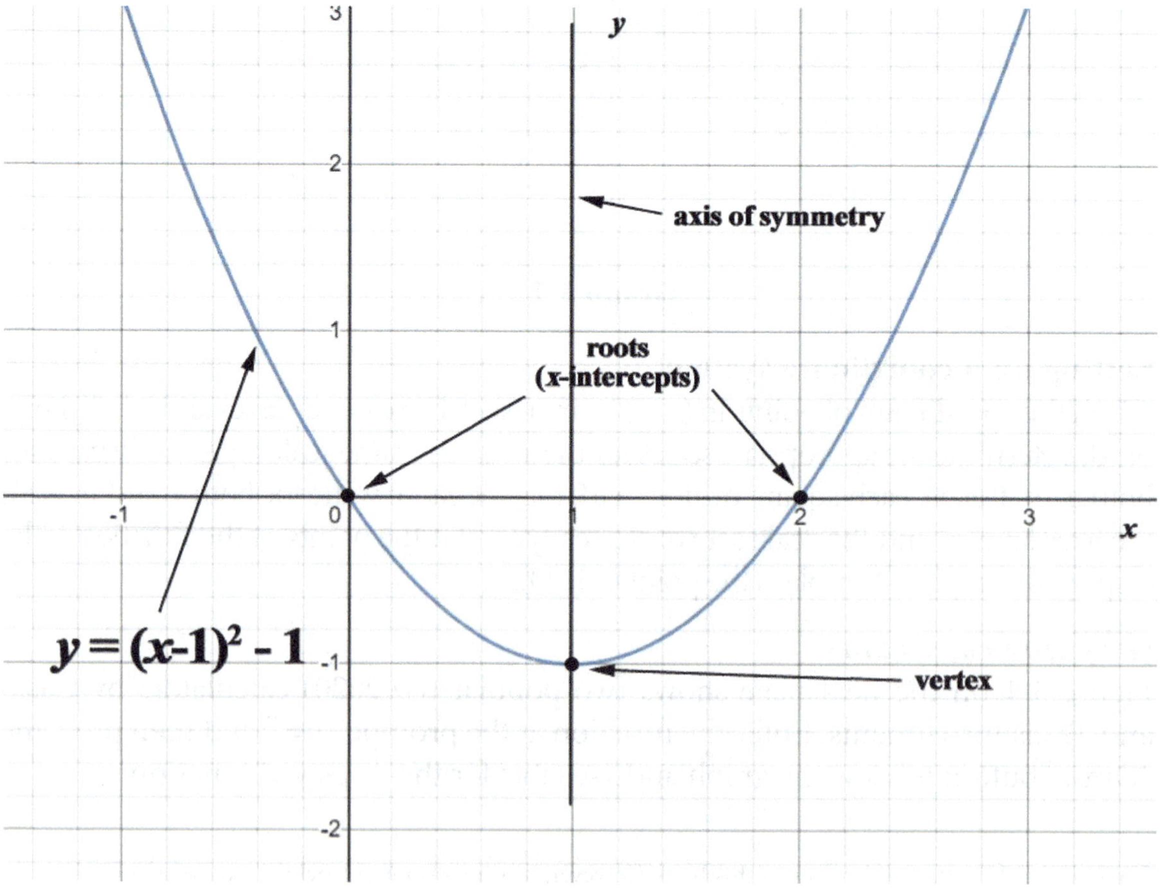

Graph 3-2

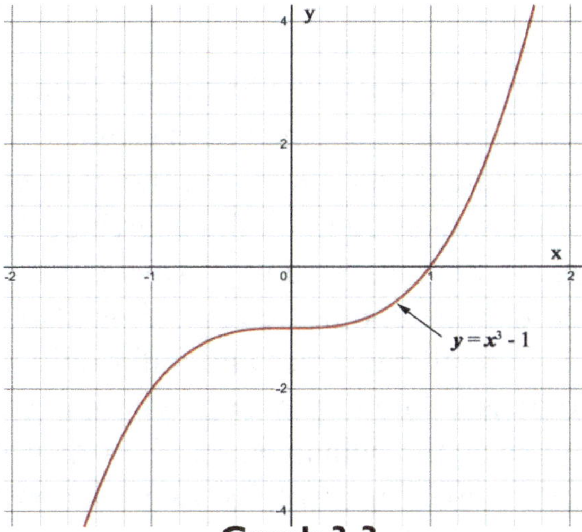

Graph 3-3

Parent functions

The functions shown in Graphs 3-2 and 3-3 are translations of the *parent functions* $y = x^2$ and $y = x^3$, respectively. These are *rigid transformations,* meaning that they do not distort the shapes. Consider the function

$$y = f(x - h) + k \qquad (3\text{-}3)$$

Equation (3-3) is a function that translates the parent function f(x) horizontally right by " h " units (if h is negative the translation is to the left) and vertically up by " k " units (if k is negative the translation is down). Graph 3-2 is a translation of the parent function x^2 horizontally to the right by 1 unit and down by one unit; Graph 3-3 is a translation of x^3 vertically down by 1 unit. Part of the analysis of a function is the identification of its parent function.

Symmetry

A function is symmetric about a vertical axis if it is identical on both sides of that axis. Look at the quadratic function graphed in Graph 3-2. One can easily vision it identical to the left and right of its *axis of symmetry* at $x = 1$. Other parent functions that are symmetric about a vertical axis include

$$f(x) = a;\ f(x) = |\,x\,|;\ f(x) = x^2 \text{ and } f(x) = \cos(x).$$

Axis of symmetry

The vertical axis is $x = 0$ for symmetric parent functions. For the quadratic in the standard form $ax^2 + bx + c = 0$, the axis of symmetry is the vertical line $x = (-b)/(2a)$.

Tests for symmetry

Given the equation of a function in x, e.g., f(x), if f(x) does not change value when $-x$ replaces x in the function, that is, if f($-x$) = f(x), then the f(x) is a *symmetrical function* and is also an *even function* as explained below.

Odd and even functions

We have seen already that an *odd function* is one where f($-x$) = $-$f(x). Odd functions are symmetric about the origin. Examples of odd functions are sin(x) and x^3. Some functions are neither odd nor even; $3x^3 - 2x^2 - x + 5$ is an example of such a function. An *even function* is one where f($-x$) = f(x). That is, if the function does not change when x is replaced by $-x$ then the function is even. The function $y = x^n + m$, where n is even and m is any real number, is an even function. Other examples* of even functions are cos(x), $|x|$, and $1/(x^2)$. Functions that are even are symmetric about vertical axes.

Roots and zeroes of a function

The *roots* (or *zeroes*) of a function are the values of x where the function becomes zero; that is, where $f(x) = 0$. If a root is a real number, then the graph crosses the x-axis st the root. Imaginary roots also exist. Graph 3-4 illustrates the concept with three quadratic functions. The number of roots is equal to the degree of the polynomial. Roots may be *real* or *imaginary;* imaginary roots always exist in *conjugate pairs*. Thus, if $(2 + 4\boldsymbol{i})$ is a root, then its conjugate $(2 - 4\boldsymbol{i})$ is also a root. The three functions shown in Graph 3-4 are all parabolas and differ only in their coordinate positions. Because a quadratic is a second-power polynomial we know that it has exactly two roots. The parabola on the right intersects the x-axis at two points. The center parabola just touches the x-axis but it still has two solutions: zero with *multiplicity* two. Finally, the parabola on the left does not touch the x-axis at all; its solutions are imaginary, namely $(-4 + \boldsymbol{i})$ and its *complex conjugate* $(-4 - \boldsymbol{i})$.

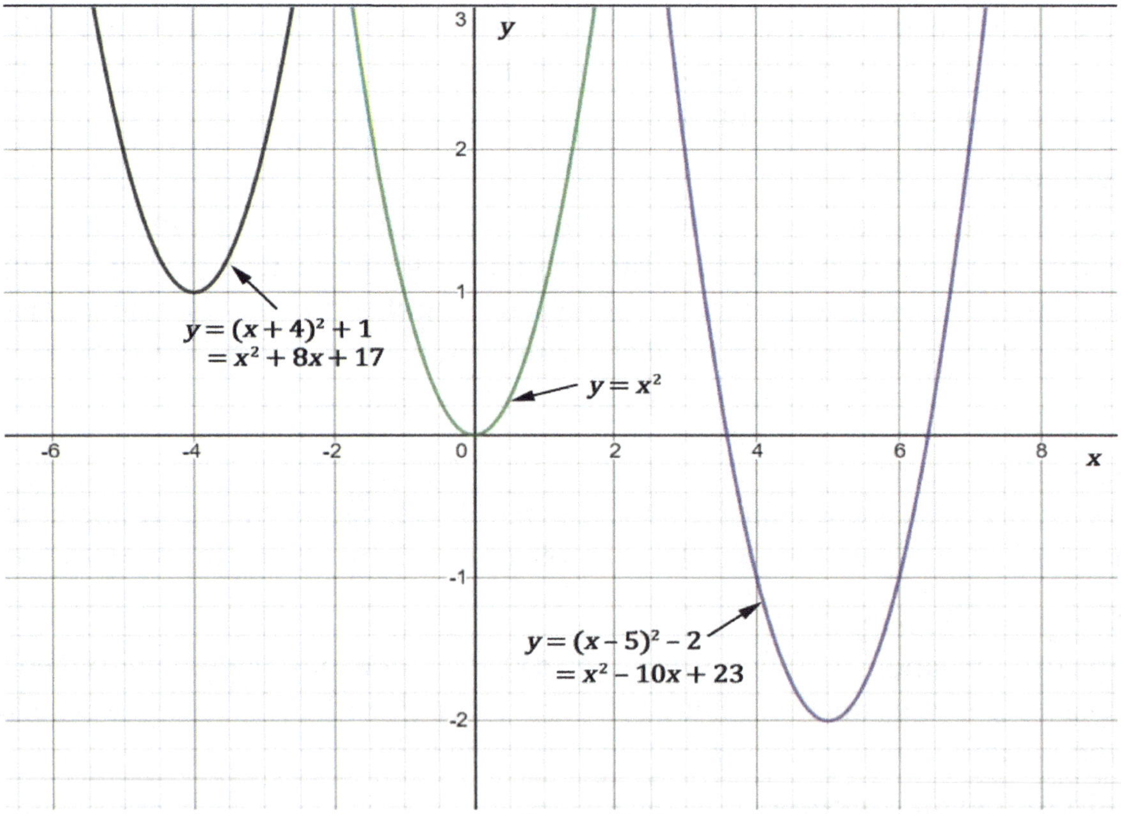

Graph 3-4

Continuous functions

Continuity

There is a test for *continuity* of a function in calculus. Intuitively, a function is continuous if it contains no "holes" or undefined values of y; that is, if its domain is $(-\infty, \infty)$. The function shown in Graph 3-5 is not continuous; it is a *rational function* (the division of one function by another) and shows a *hole*

* Complex functions (i.e., $f(z)$ where z is of the form $a + b\boldsymbol{i}$) may be even.

and two *vertical asymptotes*, i.e., values of x that are not in the domain of the function. There is also a horizontal asymptote at $y = 1$. More about this in the following discussion of asymptotes and piecewise functions.

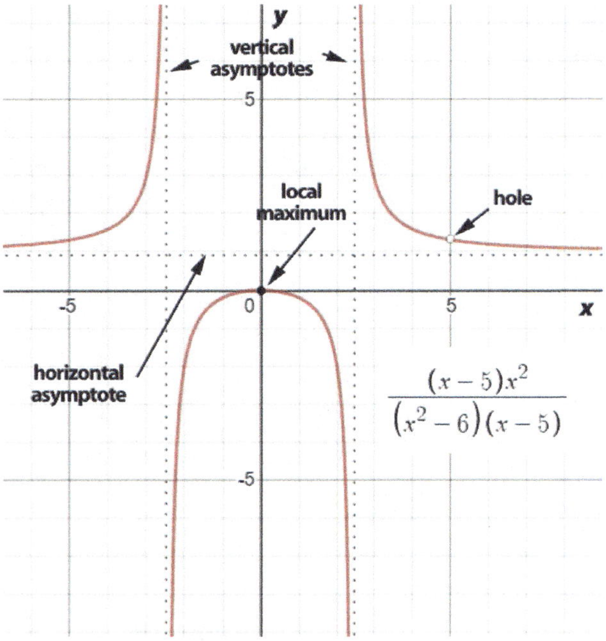

Graph 3-5

Increasing and decreasing functions
If a function f(x) increases in value as x increases, then f(x) is an *increasing function*; similarly for *decreasing*. If there are no flat intervals, then f(x) is *strictly increasing* (or *decreasing*). Generally we are interested as to whether or not a function is increasing (or decreasing) over some given interval (i.e., between two specified values of x). For example, in Graph 3-5 the function is strictly increasing in the interval on both sides of the asymptote near $x = -2$ and is strictly decreasing on both sides of the interval near $x = 2$. Note the negative part of the function is symmetric about the y-axis. The positive part is not symmetric because of the existence of the hole at $x = 5$.

Extrema and inflection points - revisited
An *extremum* (plural *extrema*) is the point at which a function has its highest (maximum) or lowest (minimum) value. A *global maximum* or *minimum* is the highest or lowest value of the entire function; a *local maximum* or *local minimum* is the highest or lowest value of f(x) in an interval within its domain. *Extrema* occur when a function changes from rising to falling or vice versa. The slope* of the function is zero at an extremum of a continuous function. Intuitively, if there is a local minimum then there must be at least one local maximum, or else the extremum would be a global *extremum* (maximum or minimum). See Graph 3-6; the slope of the function as it approaches a local minimum is decreasingly negative until at the minimum it becomes zero, after which the slope then becomes positive and increases until it approaches an inflection point, then decreases until it reaches zero at the local maximum, after which it decreases. All of this, more easily explained with differential calculus, is intuitive if you trace along the graph. The point between two local extrema where the slope changes from increasing to decreasing or vice-versa is the inflection point. Note that the slope is undefined at the point where the "hole" exists.

* This is the slope of a line that is tangent to the function at the given point.

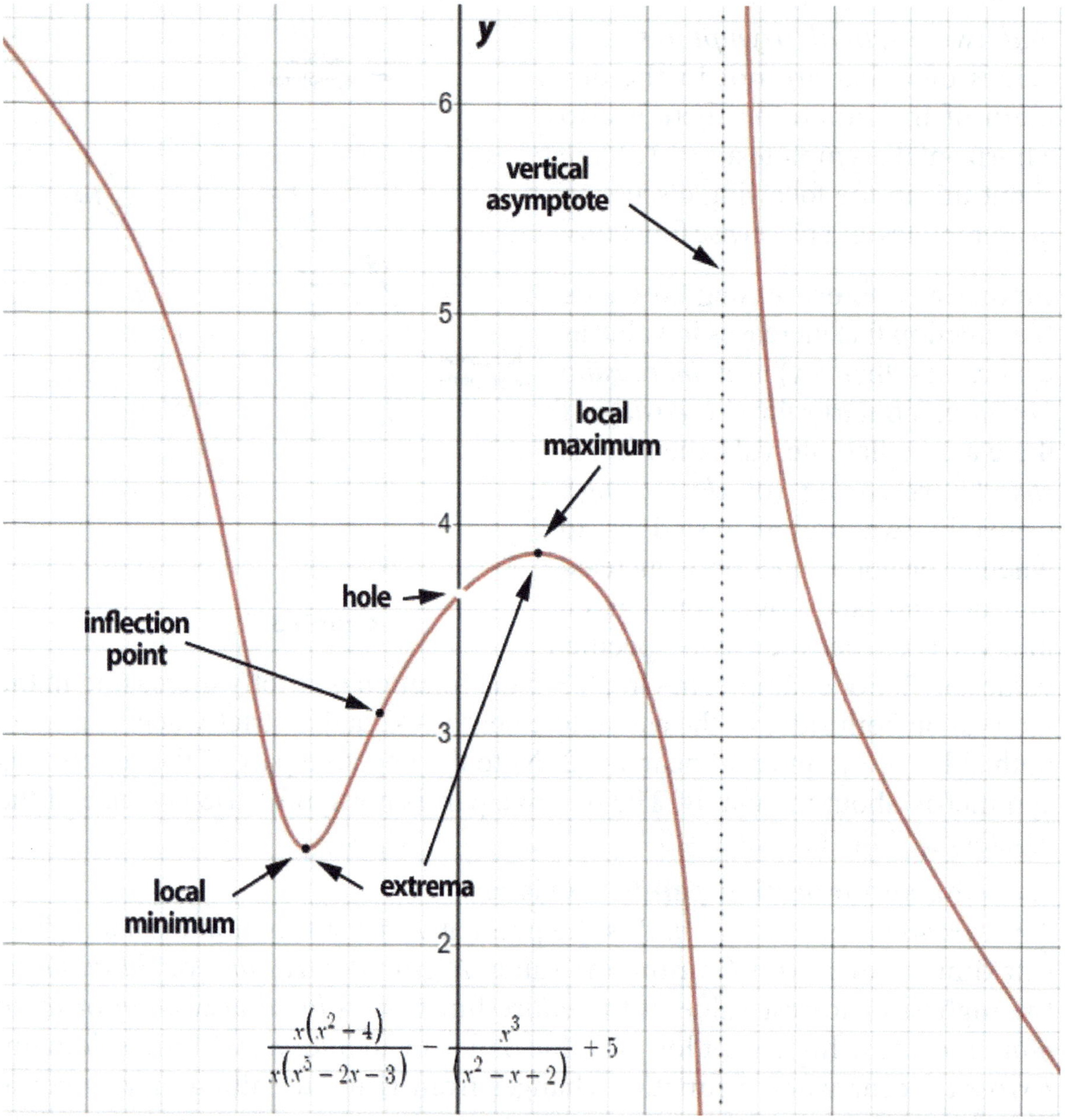

Graph 3-6

Boundedness

A function such as x^3 is unbounded; that is, it has no global extrema. The two functions in Figure 3-7 are bounded below, but have no upper bound. Note that the functions are translations of each other.

Asymptotes

If a function approaches, but does not reach, some value of y as x increases indefinitely (*end behavior*) then that y value (which is never reached *in the limit*) is a *horizontal asymptote*. If a function approaches, but does not reach, a value jof x, then that value of x is a *vertical asymptote*. It is also possible that a function approaches, but does not reach, a line $y = mx + b$. Such a line is a *slant asymptote*. See the graphs on page 3-8. It is possible for two functions to be asymptotic.

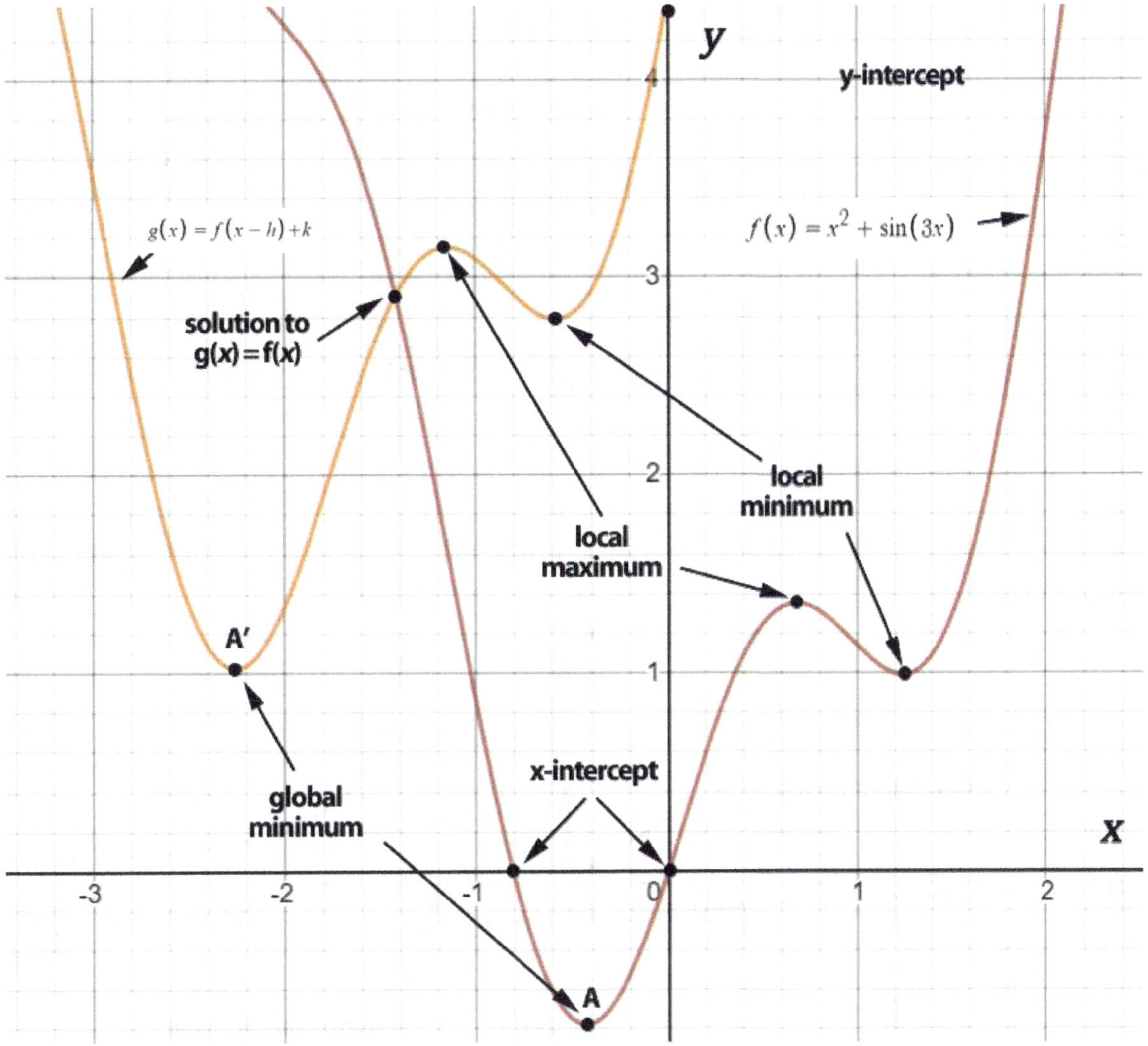

Graph 3-7

Horizontal asymptotes

Vertical asymptotes are never intersected (i.e., crossed) by a function; *horizontal asymptotes* and *slant asymptotes*, however, may be intersected once (Graph 3-9). Once crossed, a horizontal or slant asymptote is never reached again by the function in the limit (as x approaches $\pm\infty$).

Intercepts

For a given function, the points at which the function crosses or touches the x-axis are the "zeroes," or "roots," of the function. Clearly these occur when $y = 0$. Functions may intercept the y-axis; these points, where $x = 0$, are the y-intercepts of the function. Look at the functions graphed in this section for examples of x- and y-intercepts.

* Vertical asymptotes and extrema generally occur near the origin or the translated x-and y-values; end-behavior is the value that y approaches as x approaches $-\infty$ and as x approaches $+\infty$.

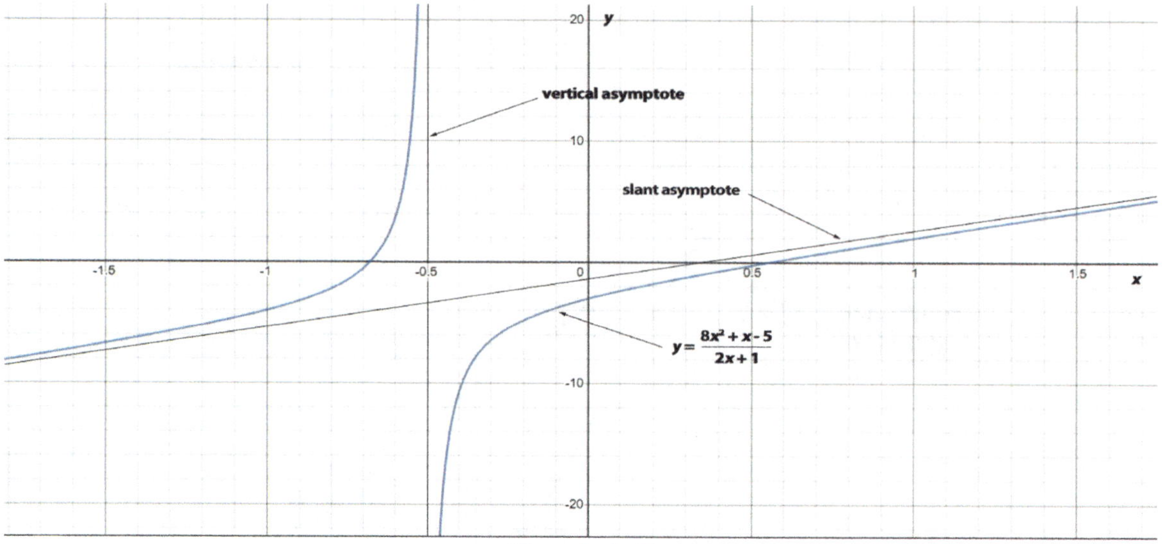

Graph 3-8

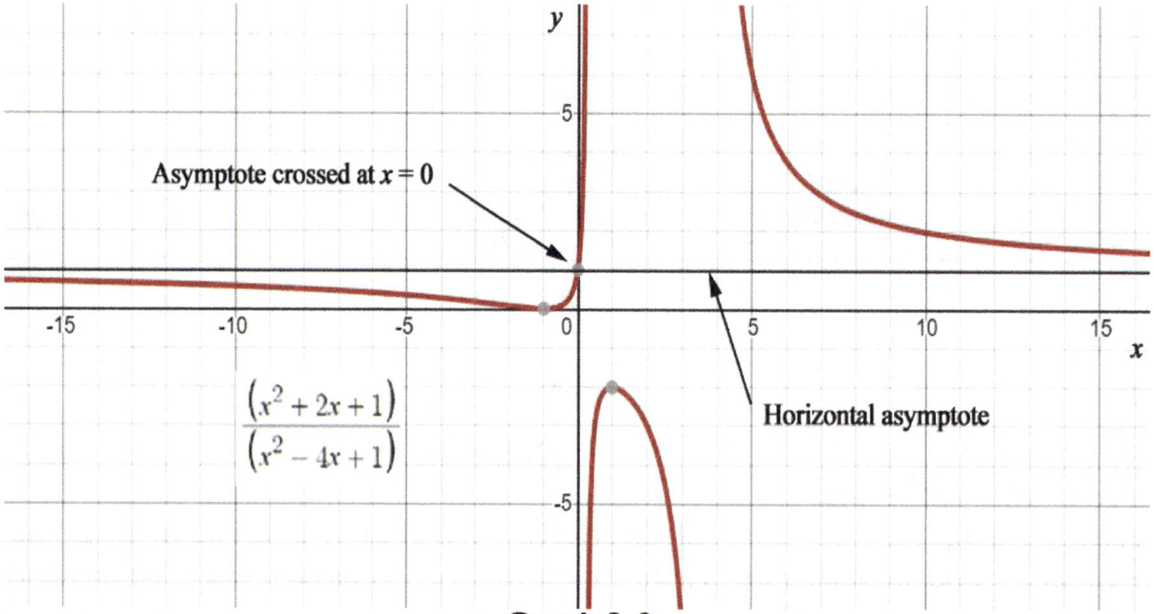

Graph 3-9

Periodicity

A function that is *periodic* repeats at fixed intervals of its domain. The function sin(x) from trigonometry is a periodic function.

Piecewise functions

If the domain of a function is partitioned into intervals and a different function is specified in each interval, then that function is a *piecewise function* such as the shown in Graph 3-10. The domain is partitioned into the three intervals (-∞, -1], (-1, 5]. and (5, ∞). Note that only one function can have a value at the end-point of an interval (else it would not be a function). There is a vertical asymptote at $x = -1$

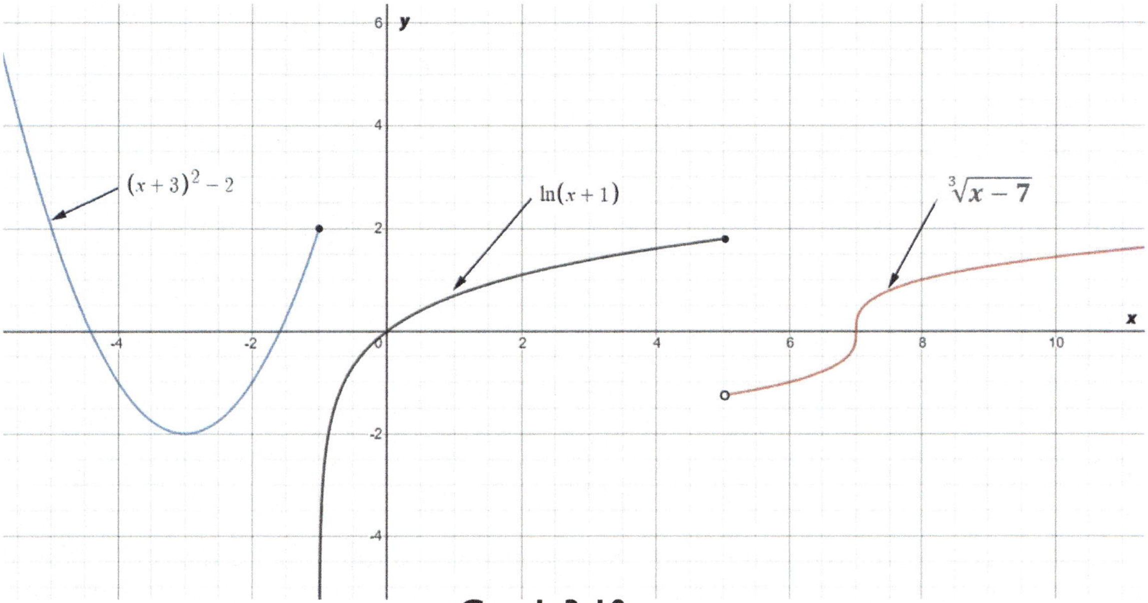

Graph 3-10

End behavior

An important consideration in the analysis of a function its behavior at extremely large values of x. That is, what does the function settle to as x approaches plus or minus infinity? Frequently in engineering, the x-axis is a domain of time (the x-variable is replaced by a t-variable). It is of interest to know how a system behaves after the initial instability when it's turned on. The range of a function becomes arbitrarily large (+ or -) as x approaches infinity, but a rational function, the ratio of two functions, say $f(x)/g(x)$, has characteristic behavior at large values* of x that depends upon the order (highest degree) of the functions in the numerator and denominator. There are three cases of interest: The degree of the polynomial (e.g., highest power of x) in the numerator is

(1) the degree of the numerator is less than the degree of the denominator: $deg[f(x)] < deg[g(x)]$;

(2) the degrees of the numerator and denominator are equal: $deg[f(x)] = deg[g(x)]$; and

(3) the degree of the numerator is equal to one plus that of the denominator: $deg[f(x)] = deg[g(x)] + 1$.

End behavior of rational functions

Case (1) - The degree of the numerator polynomial is 3 while that of the denominator is 4. As x becomes extremely large; that is, as x approaches $-\infty$ or $+\infty$, the function will begin to look like $2/x$ and y will tend to zero. Therefore, the function has a horizontal asymptote at $y = 0$ as shown in Graph 3-11.

* If the numerator has degree n, then the denominator is of degree $n - 1$.

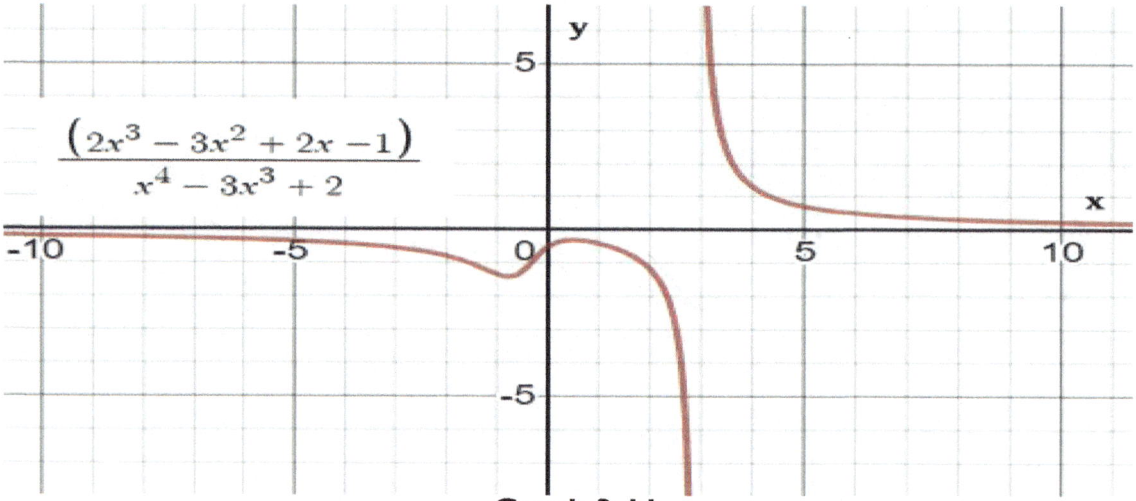

Graph 3-11

Case (2) - The degrees of the polynomials of both numerator are equal. As *x* becomes extremely large; that is, as *x* approaches -∞ and as *x* approaches +∞, the variables will cancel and the function will become a ratio of the leading coefficients which defines the horizontal asymptote; this is $y = 1/2$ for the function shown in Graph 3-12.

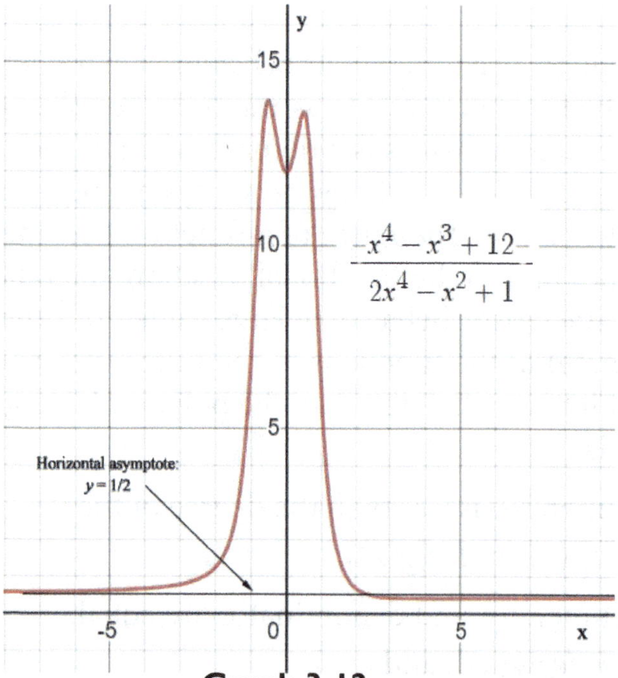

Graph 3-12

Case (3) - The degree ("4") of the numerator polynomial exceeds the degree ("3") of the denominator by unity* (i.e., 5 – 3 = 1). As *x* becomes extremely large; that is, as *x* approaches -∞ and as *x* approaches +∞, the function will begin to look like $y = x/2$ and will tend to the slant asymptote shown in Graph 3-13 on the next page. Note the intersection near $x = 2$.

Graph 3-14, on the next page, shows a function that intersects its slant asymptote.

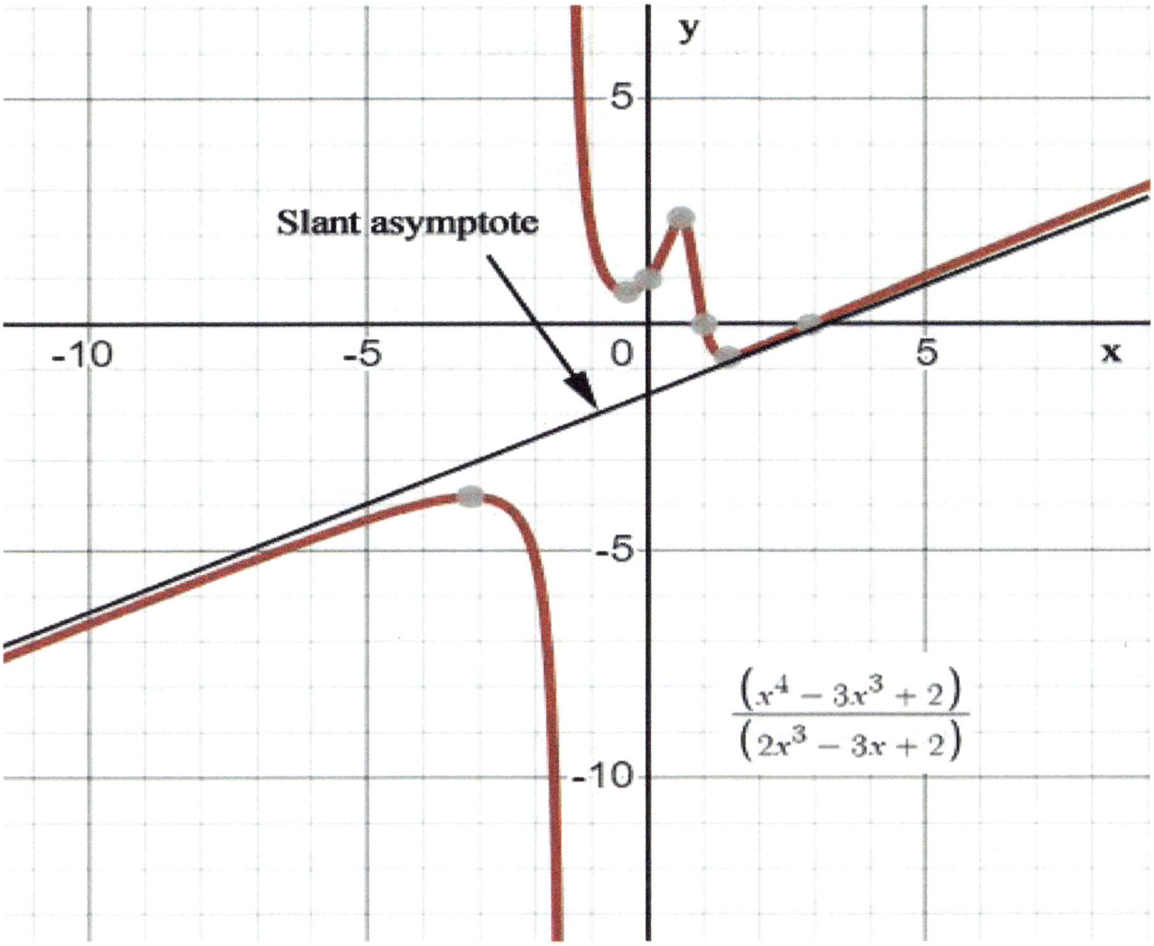

Graph 3-13

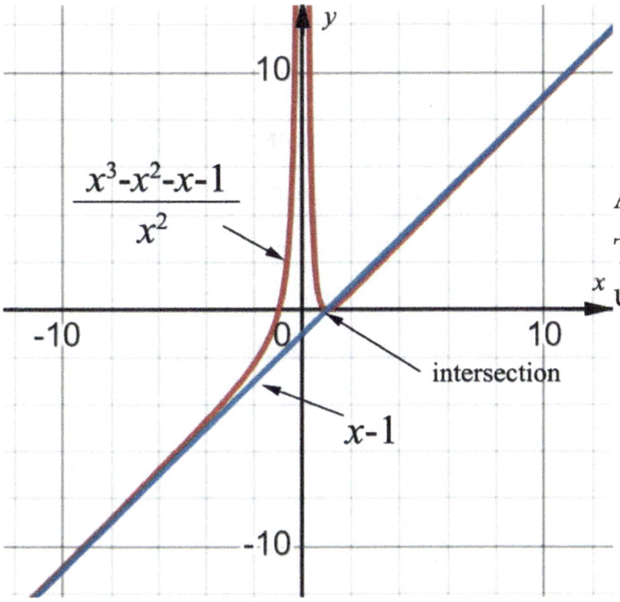

Graph 3-14

Analysis of a function worksheet

The worksheet on the next page may be useful for analyzing functions.

ANALYSIS OF THE FUNCTION

f(x) =

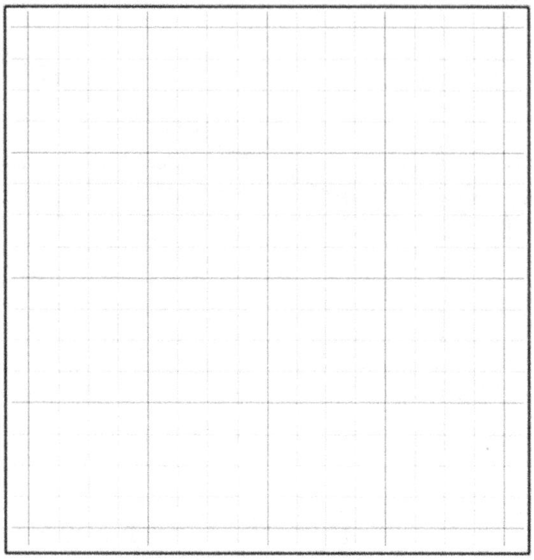

(label axes amd scales)

☐ Even function [f(x) = f(-x)]

Axis of symmetry: x =

☐ Odd function [-f(x) = f(-x)]

☐ Discontinuities: x =

☐ Holes:

☐ Vertical asymptotes:

☐ None

End Behavior

As x approaches -∞ f(x) approaches ___________

As x approaches +∞ f(x) approaches ___________

Parent Function (if any)

y = x	y = x^2
y = x^3	y = $\sqrt{x}$
y = $\lvert x \rvert$	y = $1/x$
y = a^x	y = log(x)
y = ln(x)	y = sin(x)
y = cos(x)	y = tan(x)

DOMAIN:

RANGE:

Increasing on

Decreasing on

☐ Global extrema

☐ Maximum:

☐ Minimum:

☐ Bounded above on y =

☐ Bounded below on y =

☐ Local extrema

Local maxima at (x,y):

Local minima at (x,y):

☐ Horizontal asymptote: y =

☐ Slant asymptote: y =

COMMENTS:

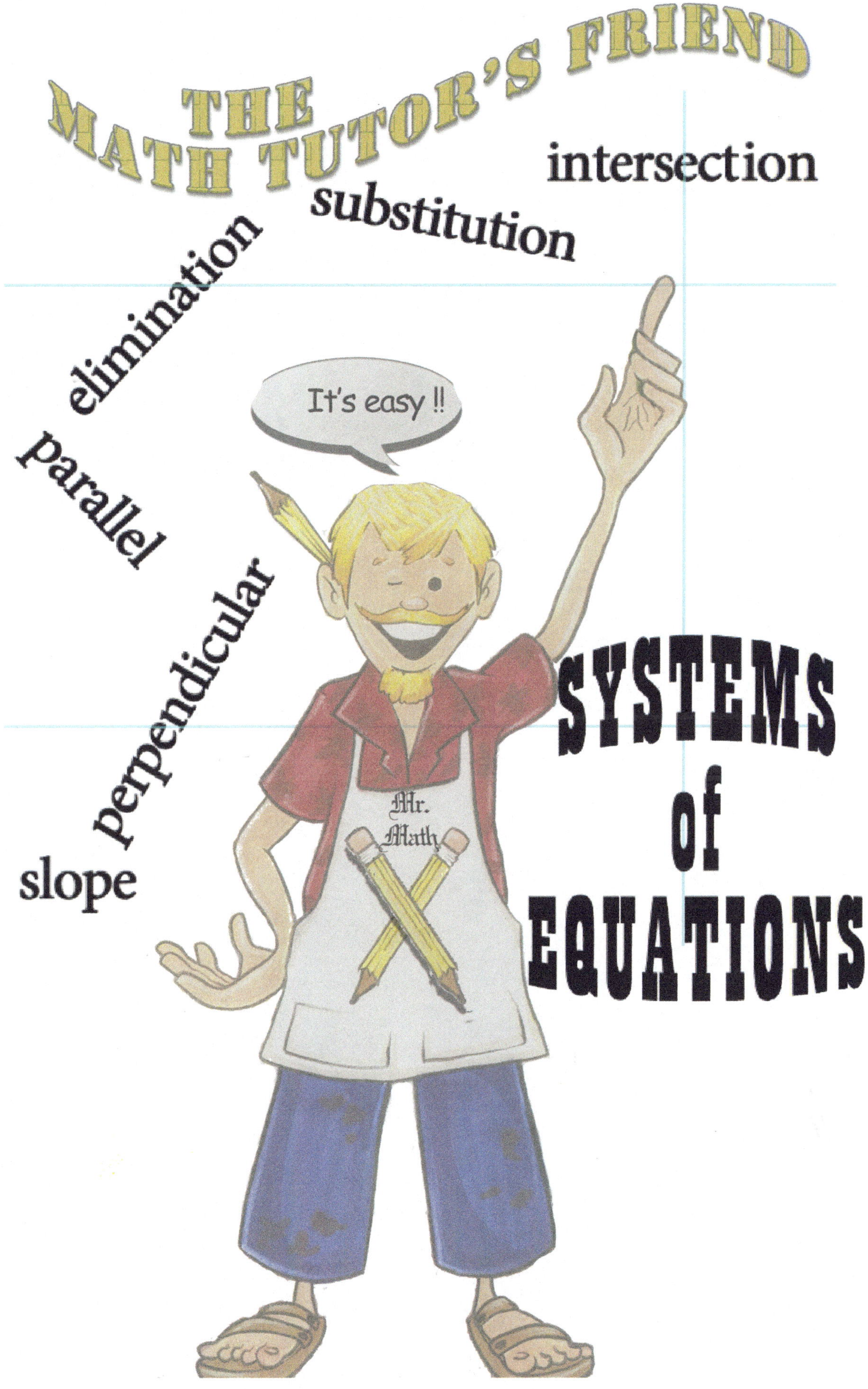
THE
MATH TUTOR'S FRIEND
intersection
substitution
elimination
It's easy !!
parallel
perpendicular
slope
Mr.
Math
SYSTEMS
of
EQUATIONS

4. SYSTEMS OF EQUATIONS
TABLE OF CONTENTS

page

•

4. SYSTEMS OF EQUATIONS

What is a linear equation?

As discussed in Section 2, a *linear equation* is the equation of a line on the x-y plane (linear equations exist in larger dimensions as intersections of planes, but we shall restrict ourselves to the 2-dimensional x-y plane where a line is defined by two points).

Systems of linear equations

If two lines are not parallel then they must intersect at some point $P(x, y)$. That is, $P(x, y)$ is a point common to both lines in question. How do we find that point? Well, there are two unknowns here: x and y, so we need two equations, that is, we need a *system of equations* to find the solution (in general, if you need to solve for three unknowns you need three equations, four equations for four unknowns, and so forth). In the case of two lines, we need to know the equation for each line to find the solution (i.e., point of intersection). Let's use the following two equations, representing two lines, and find the point of their intersection:

$$y = 3x + 5 \text{ and } y = 2x - 5$$

There are two methods for solving these equations:

(1) the *substitution method*, where one of the equations is solved for x or y and then is substituted for y or x in the other. This yields an equation with one unknown that can be solved easily. Once a value for one variable is calculated, it can be then substituted in the other equation to solve for the remaining variable. The results are the coordinates for the intersection point $P(x, y)$ which we were seeking.

(2) the *elimination method,* where the two equations are added in such a way that the term containing one of the variables is added to its *opposite* and thus eliminated (a term added to its opposite sums to zero), so then we proceed as above to solve for the other variable. Let's use both methods for the the example. We have

$$y = 3x + 5 \qquad (4\text{-}1)$$
$$y = 2x - 5 \qquad (4\text{-}2)$$

Substitution method

We can either substitute Equation (4-1) directly into Equation (4-2), or we can solve Equation (4-1) for x in terms of y and substitute it for x in Equation (4-2). Either case will result in one equation with one variable to solve. Let's do the first, that is substitute Equation (4-1) for y in Equation (4-2):

$3x + 5 = 2x - 5$ [substitute Equation (4-1) for y in Equation (4-2)]

$x + 5 = -5$ [subtract $2x$ from both sides]

$x = -10$ [subtract 5 from both sides]

"plugging" $x = -10$ into Equation (4-2) yields $y = -25$, so the lines intersect at (-10, -25)

Elimination method

To eliminate a variable, multiply one (or both) of the equations in such a way that either the x terms or the y terms (but not both) are *opposites* (additive inverse) so that when we add the two equations the x terms or the y terms will sum to zero. Looking at Equations (4-1) and (4-2) it is clear that if one of the y terms on the left side of the equation was $-y$ then they would sum to zero. So all we need do in this case is to multiply Equation (4-1) by (-1):

$$(-1)y = (-1)\,3x + (-1)\,5 \qquad \text{[Multiply Equation (4-1) by -1]}$$

$$y = 2x - 5 \qquad \text{[No change to Equation (4-2)]}$$

$$0 = -x - 10 \qquad \text{[add the two equations]}$$

We see that $x = -10$ as before, and "plugging" $x = -10$ into Equation (4-2) yields $y = -25$, as before. That was an easy example. Sometimes one equation must be multiplied by some integer or fraction and the second equation multiplied by another integer or fraction in order to achieve an additive inverse term.

Solutions to a set of linear equations

Two lines which are not parallel must intersect somewhere, and this point of intersection is found by using one of the methods just discussed. Clearly, there is exactly one point of intersection, that is, a single solution. What, then, does it mean to speak of "multiple solutions" and "no solution" when speaking of linear equations?

Multiple solutions

If two linear equations have the same slope and y- intercept, then they represent the same line and any value of x yields equal values of y. That is, the lines lie on top one another and there are an infinite number, or multiple, solutions.

$$y = 3x + 5 \qquad (4\text{-}1)$$

$$y = 3x + 5 \qquad (4\text{-}3)$$

Equations (4-1) and (4-3) have multiple (infinite) solutions.

No solution

If a system of linear equations has no solution, then the lines are parallel.

$$y = 3x + 5 \qquad (4\text{-}1)$$

$$y = 3x - 5 \qquad (4\text{-}4)$$

Equations (4-1) and (4-4) have no solution because, having the same slope and different y-intercepts, the lines are parallel and do not intersect.

Distance of a point from a line - revisited

It can be proved in geometry, and is intuitively clear, that the shortest distance from a given point P(x, y) to a line is the perpendicular distance from that line. As stated earlier, one line is perpendicular to another only if its slope is the *negative reciprocal* of the slope of the other line.

Example:

Problem: What is the distance of the point P(3, 2) to the line $y = 3x + 5$?

Solution: Point P(3, 2) must lie on a line having slope (-⅓). Substitute $x = 3$ and $y = 2$ in the slope-intercept form linear equation:

$$2 = (-\tfrac{1}{3}) \times 3 + b.$$

Solving for b yields $b = 3$. Now we have the two lines, but not yet a distance, d, of P from the original line. Let's set up the equations:

$$y_1 = 3x_1 + 5 \qquad (4\text{-}1)$$

$$y_1 = (-\tfrac{1}{3})x_1 + 3 \qquad (4\text{-}5)$$

Use the substitution method to insert Eqn (4-1) into Eqn (4-5):

$$3x_1 + 5 = (-\tfrac{1}{3})x_1 + 3$$

from which $x = -\tfrac{2}{3}$. Substituting this value of x into Equation (4-5) yields $y = 3$. Therefore, the line $y_1 = 3x_1 + 5$ intersects with its perpendicular passing through the points P(3, 2) and P(-⅔, 3). The distance, d, to be calculated is that from the point P(3, 2) to the point of intersection, P(-⅔, 3). We shall need a distance formula.

Distance Formula

We now appeal to the *distance formula*, another formula you would be wise to commit to memory:

$$d = \sqrt{(x - x_1)^2 + (y - y_1)^2}$$

The radicand (expression under the radical ("√")) is

$$3 - [-2/3])^2 + (2 - 3)^2 = (11/3)^2 + (-1)^2$$

$$= {}^{121}\!/_9 + 1 = {}^{(121 + 9)}\!/_9 = {}^{130}\!/_9.$$

Taking the square root, we have $d \approx 3.8$, to one decimal place.

Intersection of a line with a parabola

Given a line and a parabola, there are three possibilities:

1) The line intercepts the parabola at two points;
2) The line touches the parabola at a single point (tangency);
3) The line does not intercept the parabola.

Let's look at the general system of equations:

$y = mx + b$ [slope-interecept form of the equation of a line]

$y = (x - h)^2 + k$ [equation of a parabola with vertex at (h, k)]

An easy way to approximate a solution is by using a graphing calculator, or by drawing a graph.

The figure below shows a graph of the three equations

$$y = (x - 1)^2 + 2 \qquad (4\text{-}6)$$

$$y = 2x - 3 \qquad (4\text{-}7)$$

$$y = 0.5x + 3 \qquad (4\text{-}8)$$

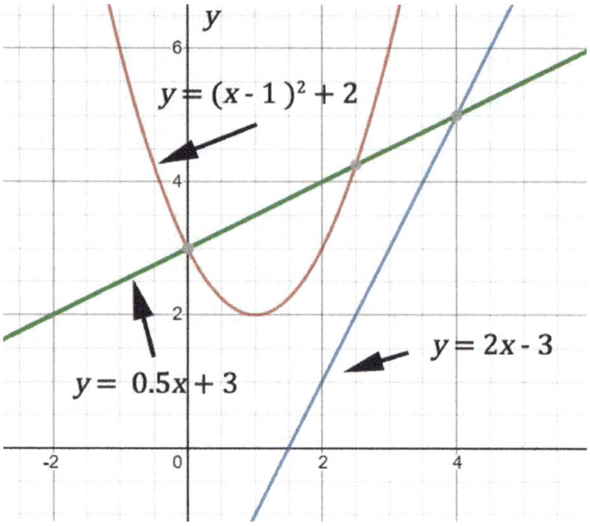

As shown on the graph, Equation (4-6) is a parabola and Equations (4-7) and (4-8) are lines. One can immediately see from the graph that the system of Equations (4-6) and (4-7) have no solution; that is, there is no point of intersection. It is clear also that Equations (4-6) and (4-8) intersect at the points (0, 3) and (2, 4.125). This is easily confirmed. Note that the two lines have different slopes so must have a point of intersection, seen on the graph at (4,5). Of course, the points of intersection can be determined by algebraic manipulation,* but it's much easier to use the graph.

A reminder about systems of equations

The solution to systems of equations are the points where they intersect; therefore, if there is no intersection there is no solution; if the equations are "on top of each other" there is an infinity of solutions.

* To determine the point(s) of intersection place each equation in the form $y = (x - 1)^2 + 2$ and $y = 0.5x + 3$ [equations (6) and (8) are already in that form] and since the point(s) of intersection have identical y-values, we can simply write

$$(x - 1)^2 + 2 = 0.5x + 3$$

then expand the left side

$x^2 - 2x + 1 + 2 = 0.5x + 3$

$x^2 - 2.5x = 0$ [subtract 3 from both sides then subtract $0.5x$ from both sides]

$x(x - 2.5) = 0$ [factor x from the left side]

The equation holds for $x = 0$ and $x = 0.25$, which is verified easily by the graph.

THE MATH TUTOR'S FRIEND

inverse

identity matrix

determinant

I_n

row operations

A×B

MATRICES & DETERMINANTS

5. MATRICES & DETERMINANTS
TABLE OF CONTENTS

•

5 - MATRICES & DETERMINANTS

Systems of linear equations

The analysis of systems of two linear equations, already discussed, can be expanded to three or more equations. When three or more linear variables are involved, it may be simpler to employ a *matrix* to solve the system of linear equations.

Matrices (plural of "matrix")

A *matrix* is simply a set of numbers, letters, or other objects arranged into rows and columns – i.e., a rectangular array. For systems of equations, the *coefficients* of the variables form a matrix which is employed to solve the given system of equations. Let's use bold-face capital letters to represent matrices.

$$\mathbf{A} = \begin{pmatrix} a_{11} & a_{12} & a_{13} & a_{14} \\ a_{21} & a_{22} & a_{23} & a_{24} \\ a_{31} & a_{32} & a_{33} & a_{34} \\ a_{41} & a_{42} & a_{43} & a_{44} \end{pmatrix}$$

For example, let Matrix **A** be a 4×4 ("4-by-4") *two-dimensional matrix*:* that is, an array of four rows and four columns. In general, a matrix with m rows and n columns is referred to as a matrix of *order* $m \times n$**. Its *main diagonal* consists of the *elements* a_{11}, a_{22}, a_{33}, a_{44}. The two *subscripts* ij identify element ij in a matrix; that is ij refers to row i and column j of the matrix. For example, element a_{43} is the entry in the fourth row and third column of matrix **A**. Matrix **A** in the example is a *square matrix*, which has order $m \times m$ or "m by m."

Other ways of denoting a matrix are to use the expression $\mathbf{A} = [a_{ij}]$ or sometimes $\mathbf{A}_{ij}$. Note that the order (or dimension) of a row in a matrix is a $1 \times n$ matrix called a *row vector*. Likewise, a *column vector* is an $m \times 1$ matrix (i.e., m rows in one column). Finally, a 1×1 matrix, which is just a number, is a *scalar*.

Square matrix

A square matrix has dimension $m \times m$, that is, the number of rows is equal to the number of columns.

* One may envision three-dimension matrices as a "stack" of two-dimensional matrices with identical numbers of rows and columns. Elements would have three subscripts to identify their locations in the matrix.

** The symbol "×" means "by" in the context of the order of a matrix; thus, $m \times n$ means "m by n" and refers to the order or dimension of the matrix – that is, m rows and n columns. An *m* by *n matrix has mn elements, so in that sense m×n is a product.* There is another interpretation in connection with vectors where "×" means "cross product" and is beyond the scope of this discussion.

Diagonal matrix

A square matrix for which all of the elements not on the main diagonal are zero is a *diagonal matrix.* At least one element on the main diagonal must be non-zero. Matrix **D** is a diagonal matrix.

$$\mathbf{D} = \begin{pmatrix} a_{11} & 0 & 0 & 0 \\ 0 & a_{22} & 0 & 0 \\ 0 & 0 & a_{33} & 0 \\ 0 & 0 & 0 & a_{44} \end{pmatrix}$$

Trace of a matrix

The *trace of a matrix* **A**, denoted by tr(**A**), is the sum of the elements in the main diagonal of the matrix. Note that every element in the main diagonal has matching subscripts; that is, for each a_{ij} on the main diagonal, $i = j$.

Triangular forms of a square matrix

Matrices **A** and **B** shown below are examples of the *triangular form.* Matrix **A**, with zeroes below the diagonal, is in *upper triangular form* and matrix **B** is in *lower triangular form.*

$$\mathbf{A} = \begin{pmatrix} 1 & 3 & 0 & 2 \\ 0 & 5 & 5 & 7 \\ 0 & 0 & 4 & 3 \\ 0 & 0 & 0 & 9 \end{pmatrix} \qquad \mathbf{B} = \begin{pmatrix} 1 & 0 & 0 & 0 \\ 6 & 4 & 0 & 0 \\ 5 & 9 & 5 & 0 \\ 3 & 5 & 7 & 7 \end{pmatrix}$$

Identity matrix

A diagonal matrix in which all elements on the main diagonal are "1" and the remaining elements "0," is an *identity matrix*. The $n \times n$ identity matrix is usually symbolized by $\mathbf{I}_n$. Note that $\mathrm{tr}(\mathbf{I}_n) = n$.

$$\mathbf{I}_4 = \begin{pmatrix} 1 & 0 & 0 & 0 \\ 0 & 1 & 0 & 0 \\ 0 & 0 & 1 & 0 \\ 0 & 0 & 0 & 1 \end{pmatrix}$$

Transpose of a matrix

Let **A** be an $m \times n$ matrix. Then the *transpose* of **A**, signified by $\mathbf{A}^T$, is an $n \times m$ matrix in which the rows of **A** become the columns of $\mathbf{A}^T$ and the columns of **A** become the rows of $\mathbf{A}^T$. See the examples below.

$$\mathbf{A} = \begin{pmatrix} 1 & 2 & 3 \\ 4 & 5 & 6 \\ 7 & 8 & 9 \end{pmatrix} \qquad \mathbf{A}^T = \begin{pmatrix} 1 & 4 & 7 \\ 2 & 5 & 8 \\ 3 & 6 & 9 \end{pmatrix}$$

Adding matrices

Matrices may be added (or subtracted) only if they have the same number of rows and columns i.e., the same order $m \times n$. An example showing the addition of two 3×3 matrices A and B appears below. The sum/difference of two $m \times n$ matrices is an $m \times n$ matrix consisting of elements that are the sums/differences of the corresponding elements of the original matrices, as shown.

$$\mathrm{A} = \begin{pmatrix} a_{11} & a_{12} & a_{13} \\ a_{21} & a_{22} & a_{23} \\ a_{31} & a_{32} & a_{33} \end{pmatrix}$$

$$\mathrm{B} = \begin{pmatrix} b_{11} & b_{12} & b_{13} \\ b_{21} & b_{22} & b_{23} \\ b_{31} & b_{32} & b_{33} \end{pmatrix}$$

$$\mathrm{A} + \mathrm{B} = \begin{pmatrix} a_{11} & a_{12} & a_{13} \\ a_{21} & a_{22} & a_{23} \\ a_{31} & a_{32} & a_{33} \end{pmatrix} + \begin{pmatrix} b_{11} & b_{12} & b_{13} \\ b_{21} & b_{22} & b_{23} \\ b_{31} & b_{32} & b_{33} \end{pmatrix} = \begin{pmatrix} a_{11} + b_{11} & a_{12} + b_{12} & a_{13} + b_{13} \\ a_{21} + b_{21} & a_{22} + b_{22} & a_{23} + b_{23} \\ a_{31} + b_{31} & a_{32} + b_{32} & a_{33} + b_{33} \end{pmatrix}$$

Product of a real number and a matrix

Let k be a real number. The product of k with a matrix $\mathbf{A} = [a_{ij}]$ is simply $[ka_{ij}]$. Multiply each element of **A** by the real number k.

Product of two matrices

Dot product

When dealing with vectors* there is a multiplication technique called the *dot product* defined as follows: The dot product of row i vector $a_i = a_{i1}, a_{i2}, a_{i3}, \ldots, a_{im})$ and column j vector $b_j = (b_{1j}, b_{2j}, b_{3j}, \ldots, b_{nj})^T$

Two matrices may be multiplied only if they are in *conformance*; that is, an $m \times n$ matrix **A** is *conformable for multiplication* with a matrix **B** if and only if **B** has order $n \times r$. That is the number of columns in **A** must be the same as the number of rows in **B**. The product will be an $m \times r$ matrix.

$$\mathbf{A} = \begin{pmatrix} a_{11} & a_{12} \\ a_{21} & a_{22} \\ a_{31} & a_{32} \end{pmatrix}$$

$$\mathbf{B} = \begin{pmatrix} b_{11} & b_{12} & b_{13} \\ b_{21} & b_{22} & b_{23} \end{pmatrix}$$

* The vectors of interest when dealing with matrices are the row vector and the column vector, already defined.

$$(a_{i1}\ a_{i2}, \ldots, a_{in}) \cdot \begin{pmatrix} b_{1j} \\ b_{2j} \\ \vdots \\ b_{nj} \end{pmatrix} = a_{i1}b_{1j} + a_{i2}b_{2j} + \ldots + a_{in}b_{nj}$$

$$\text{or } a_i \cdot b_j = \sum_{k=1}^{n} a_{ik}\, b_{kj}$$

The order of **A** is 3×2 and that of **B** is 2×3. The product of the 3×2 matrix with the 2×3 matrix will be a 3×3 matrix; call it **C**.

$$\mathbf{C} = \begin{pmatrix} c_{11} & c_{12} & c_{13} \\ c_{21} & c_{22} & c_{23} \\ c_{31} & c_{32} & c_{33} \end{pmatrix}$$

In general, for the matrix multiplication **C** = **AB**, element c_{ij} is the dot product of row *i* of matrix **A** with column *j* of matrix **B**. The product matrix **C** is formed from matrices **A** and **B** as follows:

$$\text{AB} = \begin{pmatrix} (a_{11}b_{11} + a_{12}b_{21}) & (a_{11}b_{12} + a_{12}b_{22}) & (a_{11}b_{13} + a_{12}b_{23}) \\ (a_{21}b_{11} + a_{22}b_{21}) & (a_{21}b_{12} + a_{22}b_{22}) & (a_{21}b_{13} + a_{22}b_{23}) \\ (a_{31}b_{11} + a_{32}b_{21}) & (a_{31}b_{12} + a_{32}b_{22}) & (a_{31}b_{13} + a_{32}b_{23}) \end{pmatrix}$$

Multiplication of matrices is not *commutative*; that is, **AB** ≠ **BA**; **A** and **B** are inverses of each other.

Determinant of a (square) matrix

Consider the 2×2 matrix **A** shown below. Its four elements are *a, b, c,* and *d.* The determinant, det(**A**), of **A** is indicated by vertical bars instead of parentheses or brackets.

$$\mathbf{A} = \begin{pmatrix} a & b \\ c & d \end{pmatrix} \qquad \det(\mathbf{A}) = \begin{vmatrix} a & b \\ c & d \end{vmatrix} = ad - bc$$

$c_{11} = (a_{11}\ a_{12}) \cdot \begin{pmatrix} b_{11} \\ b_{21} \end{pmatrix} = (a_{11}b_{11} + a_{12}b_{21})$ Row 1 of A times Column 1 of B

$c_{12} = (a_{11}\ a_{12}) \cdot \begin{pmatrix} b_{12} \\ b_{22} \end{pmatrix} = (a_{11}b_{12} + a_{12}b_{22})$ Row 1 of A times Column 2 of B

$c_{13} = (a_{11}\ a_{12}) \cdot \begin{pmatrix} b_{13} \\ b_{23} \end{pmatrix} = (a_{11}b_{13} + a_{12}b_{23})$ Row 1 of A times Column 3 of B

$c_{21} = (a_{21}\ a_{22}) \cdot \begin{pmatrix} b_{11} \\ b_{21} \end{pmatrix} = (a_{21}b_{11} + a_{22}b_{21})$ Row 2 of A times Column 1 of B

$c_{22} = (a_{21}\ a_{22}) \cdot \begin{pmatrix} b_{12} \\ b_{22} \end{pmatrix} = (a_{21}b_{12} + a_{22}b_{22})$ Row 2 of A times Column 2 of B

$c_{23} = (a_{21}\ a_{22}) \cdot \begin{pmatrix} b_{13} \\ b_{23} \end{pmatrix} = (a_{21}b_{13} + a_{22}b_{23})$ Row 2 of A times Column 3 of B

$c_{31} = (a_{31}\ a_{32}) \cdot \begin{pmatrix} b_{11} \\ b_{21} \end{pmatrix} = (a_{31}b_{11} + a_{32}b_{21})$ Row 3 of A times Column 1 of B

$c_{32} = (a_{31}\ a_{32}) \cdot \begin{pmatrix} b_{12} \\ b_{22} \end{pmatrix} = (a_{31}b_{12} + a_{32}b_{22})$ Row 3 of A times Column 2 of B

$c_{33} = (a_{31}\ a_{32}) \cdot \begin{pmatrix} b_{13} \\ b_{23} \end{pmatrix} = (a_{31}b_{13} + a_{32}b_{23})$ Row 3 of A times Column 3 of B

If $ad = bc$ then det(**A**) = 0 and **A** is a *singular matrix.*

Examples: Consider the two square matrices **A** and **B** shown below. Calculate the determinant of each of these; that is, calculate det(**A**) and det(**B**). Note that matrix **A** is *non-singular* and matrix **B** is singular.

$$\mathbf{A} = \begin{pmatrix} 2 & -8 \\ 1 & 4 \end{pmatrix} \qquad \det(\mathbf{A}) = \begin{vmatrix} 2 & -8 \\ 1 & 4 \end{vmatrix} = 2\cdot4 - (-8)\cdot1 = 16$$

$$\mathbf{B} = \begin{pmatrix} 2 & 8 \\ 1 & 4 \end{pmatrix} \qquad \det(\mathbf{B}) = \begin{vmatrix} 2 & 8 \\ 1 & 4 \end{vmatrix} = 2\cdot4 - 8\cdot1 = 0$$

Determinant of a 3×3 matrix

How about higher-order square matrices such as 3×3, 4×4, *etc*? Before discussing this we introduce the term *minor* (or *minor determinant*). Look at the 3×3 matrix **A** below.*

$$\mathbf{A} = \begin{pmatrix} a_{11} & a_{12} & a_{13} \\ a_{21} & a_{22} & a_{23} \\ a_{31} & a_{32} & a_{33} \end{pmatrix} = \begin{pmatrix} 2 & 6 & 2 \\ 4 & 5 & 1 \\ 3 & 0 & 1 \end{pmatrix}$$

Select column 1 of **A** (any other column may be selected) which has entries $a_{11} = 2$, $a_{21} = 4$ and $a_{31} = 3$. Now choose element $a_{11} = 2$. Mentally delete the row and column in which $a_{11} = 2$ appears to display the 2×2 minor determinant (or minor)

$$\mathbf{M}_{11} = \begin{vmatrix} 5 & 1 \\ 0 & 1 \end{vmatrix}$$

which evaluates to 5 – 0 = 5, the *cofactor* C_{11} of a_{11}.

Next, select $a_{21} = 4$. Delete row 2 and column 1 to display the minor

$$\mathbf{M}_{21} = \begin{vmatrix} 6 & 2 \\ 0 & 1 \end{vmatrix}$$

which evaluates to 6 – 0 = 6, the cofactor C_{21} of a_{21}.

Finally, element $a_{31} = 3$ has minor $\mathbf{M}_{31} = \begin{vmatrix} 6 & 2 \\ 5 & 1 \end{vmatrix}$

which evaluates 6 – 10 = -4, the cofactor C_{31} of a_{31}.

Now evaluate the determinant of the 3×3 matrix **A** by multiplying each element column 1 by its minor calculated above and adding them according to the rule "if the sum of the subscripts *ij* of the element is odd then precede the product with a minus sign otherwise use a plus sign." This means simply that the a_{ij} is multiplied by $(-1)^{i+j}$. In this example the sum of the subscript digits of c_{21} is odd, so a minus sign precedes the term 4·6.

$$\begin{vmatrix} 2 & 6 & 2 \\ 4 & 5 & 1 \\ 3 & 0 & 1 \end{vmatrix} = 2\cdot5 - 4\cdot6 + 3\cdot(-4) = -26$$

* Determinants apply only to square matrices, i.e., matrices of order *m*×*m*.

Alternate method for computing the determinant

This method is a "shortcut" to compute a determinant of a square matrix without resorting to minors. In our 3×3 example of det(**A**) copy the first two columns to the right of the determinant array.

$$\begin{vmatrix} 2 & 6 & 2 \\ 4 & 5 & 1 \\ 3 & 0 & 1 \end{vmatrix} \begin{matrix} 2 & 6 \\ 4 & 5 \\ 3 & 0 \end{matrix}$$

Next, draw or visualize the diagonals as shown below.

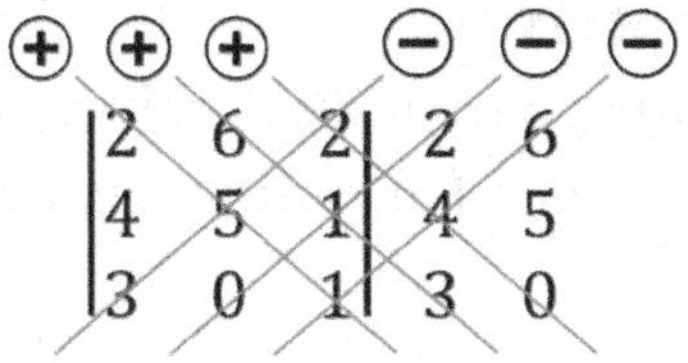

Now sum the products of the entries along each diagonal, adding or subtracting according to the signs in the circles:

$$10 + 18 + 0 - 24 - 0 - 30 = -26$$

is the value of det(**A**) as already calculated using the previous method of minors.

Submatrix

If one deletes an identical number of rows and columns from a matrix **A**, the remaining elements form a *submatrix*. The minors we worked with above are submatrices.

Cofactor

As we have seen, the minor for element e_{ij} in a matrix is a submatrix formed by deleting row i and column j at the location of element e_{ij}. The determinant, M_{ij}, of the minor multiplied by $(-1)^{i+j}$ is called the (i, j) *cofactor* C_{ij}. For example, consider the 3×3 matrix **A** below:

$$\mathbf{A} = \begin{pmatrix} 1 & 2 & 0 \\ 4 & 5 & 6 \\ 7 & 8 & 9 \end{pmatrix}$$

The determinant of **A** is calculated from the augmented array

$$\begin{vmatrix} 1 & 2 & 0 \\ 4 & 5 & 6 \\ 7 & 8 & 9 \end{vmatrix} \begin{matrix} 1 & 2 \\ 4 & 5 \\ 7 & 8 \end{matrix}$$

$$\det(\mathbf{A}) = 45 + 84 + 0 - 72 - 48 - 0 = 9$$

The minor, M_{21}, of element a_{21} is

$$\det\begin{pmatrix} \square & 2 & 0 \\ \square & \square & \square \\ \square & 8 & 9 \end{pmatrix} = \begin{vmatrix} 2 & 0 \\ 8 & 9 \end{vmatrix} = 18 - 0 = 18$$

and the cofactor, C_{21}, of element a_{21} is $= (-1)\cdot M_{21} = -18$.

Matrix of minors

For the example,the matrix,M_A, of minors for A is

$$\mathbf{M_A} = \begin{pmatrix} \mathbf{M_{11}} & \mathbf{M_{12}} & \mathbf{M_{13}} \\ \mathbf{M_{21}} & \mathbf{M_{22}} & \mathbf{M_{23}} \\ \mathbf{M_{31}} & \mathbf{M_{32}} & \mathbf{M_{33}} \end{pmatrix}$$

$$= \begin{pmatrix} \det\begin{pmatrix}5 & 6\\8 & 9\end{pmatrix} & \det\begin{pmatrix}4 & 6\\7 & 9\end{pmatrix} & \det\begin{pmatrix}4 & 5\\7 & 8\end{pmatrix} \\ \det\begin{pmatrix}2 & 0\\8 & 9\end{pmatrix} & \det\begin{pmatrix}1 & 0\\7 & 9\end{pmatrix} & \det\begin{pmatrix}1 & 2\\7 & 8\end{pmatrix} \\ \det\begin{pmatrix}2 & 0\\5 & 6\end{pmatrix} & \det\begin{pmatrix}1 & 0\\4 & 6\end{pmatrix} & \det\begin{pmatrix}1 & 2\\4 & 5\end{pmatrix} \end{pmatrix} = \begin{pmatrix} \begin{vmatrix}5 & 6\\8 & 9\end{vmatrix} & \begin{vmatrix}4 & 6\\7 & 9\end{vmatrix} & \begin{vmatrix}4 & 5\\7 & 8\end{vmatrix} \\ \begin{vmatrix}2 & 0\\8 & 9\end{vmatrix} & \begin{vmatrix}1 & 0\\7 & 9\end{vmatrix} & \begin{vmatrix}1 & 2\\7 & 8\end{vmatrix} \\ \begin{vmatrix}2 & 0\\5 & 6\end{vmatrix} & \begin{vmatrix}1 & 0\\4 & 6\end{vmatrix} & \begin{vmatrix}1 & 2\\4 & 5\end{vmatrix} \end{pmatrix}$$

$$\mathbf{M_A} = \begin{pmatrix} -3 & -6 & -3 \\ 18 & 9 & -6 \\ 12 & 6 & -3 \end{pmatrix}$$

Cofactor matrix

Recall that the cofactor of minor determinant M_{ij} is $C_{ij} = (-1)^{i+j} \cdot M_{ij}$. Thus, the *cofactor matrix*, C_A, for matrix **A** is

$$\mathbf{C_A} = \begin{pmatrix} -3 & 6 & -3 \\ -18 & 9 & 6 \\ 12 & -6 & -3 \end{pmatrix}$$

Adjugate of an n×n matrix

The *adjugate* (*adjunct*) of a square matrix is the transpose of its cofactor matrix. For the example,

$$\text{adj}(\mathbf{C_A}) = (\mathbf{C^T})_A = \begin{pmatrix} -3 & -18 & 12 \\ 6 & 9 & -6 \\ -3 & 6 & -3 \end{pmatrix}$$

Inverse of a matrix

Invertible matrix

An $n \times n$ square matrix **A** is invertible* if there exists another matrix **B** of the same order such that $\mathbf{AB} = \mathbf{I}_n$ where $\mathbf{I}_n$ is the identity matrix. Matrix **B** is the inverse of matrix **A** and is symbolized by $\mathbf{A}^{-1}$. Thus, $\mathbf{AA}^{-1} = \mathbf{I}_n$.

In summary, matrix inversion of **A** is accomplished by finding a matrix **B** that satisfies the equation $\mathbf{AB} = \mathbf{I}_n$.

But just how do you calculate a matrix inverse? First let's talk about *row operations* on a matrix. We already have discussed the computations necessary to calculate the inverse. In short, the inverse of invertible square matrix A is given by

$$\mathbf{A}^{-1} = \frac{1}{\det(\mathbf{A})} \cdot \text{adj}(\mathbf{A})$$

* If the determinant of a matrix is not zero, then that matrix is invertible; if the determinant is zero then it is noninvertible (also singular or degenerate)

For the example,

$$\mathbf{A}^{-1} = \frac{1}{9}\begin{pmatrix} -3 & -18 & 12 \\ 6 & 9 & -6 \\ -3 & 6 & -3 \end{pmatrix} = \begin{pmatrix} -1/3 & -2 & 4/3 \\ 2/3 & 1 & -2/3 \\ -1/3 & 2/3 & -1/3 \end{pmatrix}$$

We can verify this by showing that $\mathbf{AA}^{-1} = \mathbf{I}_3$

Another technique for computing the inverse of a matrix

Find the inverse of the following matrix.*

$$\begin{pmatrix} \mathbf{1} & \mathbf{3} & \mathbf{3} \\ \mathbf{1} & \mathbf{4} & \mathbf{3} \\ \mathbf{1} & \mathbf{3} & \mathbf{4} \end{pmatrix}$$

The rows in the augmented matrix from top to bottom are R1, R2, and R3.

Row operations

Row operations, or elementary row operations, convert an augmented matrix to another augmented matrix in which the identity matrix will appear on the left side of the vertical line. The row operations are:

- Interchange any two rows –
  Example, "Interchange R1 and R3."
- Multiply elements of a row by a nonzero real number
  Example, "-3R1→ R1 (multiply R1 by -3),
- Add a multiple of one row to another row
  Example, "R2 → R1 + R2."

The first step is to augment the matrix with the identity matrix $\mathbf{I}_3$:

$$\left(\begin{array}{ccc|ccc} 1 & 3 & 3 & 1 & 0 & 0 \\ 1 & 4 & 3 & 0 & 1 & 0 \\ 1 & 3 & 4 & 0 & 0 & 1 \end{array}\right)$$

Next, perform the first and second row operations to get the matrix on the right.

$$\left(\begin{array}{ccc|ccc} 1 & 3 & 3 & 1 & 0 & 0 \\ 1 & 4 & 3 & 0 & 1 & 0 \\ 1 & 3 & 4 & 0 & 0 & 1 \end{array}\right) \xrightarrow[\text{-R1 + R3}]{\text{-R1 + R2}} \left(\begin{array}{ccc|ccc} 1 & 3 & 3 & 1 & 0 & 0 \\ 0 & 1 & 0 & -1 & 1 & 0 \\ 0 & 0 & 1 & -1 & 0 & 1 \end{array}\right)$$

$$\xrightarrow{\text{-3R2 + R1}} \left(\begin{array}{ccc|ccc} 1 & 0 & 3 & 4 & -3 & 0 \\ 0 & 1 & 0 & -1 & 1 & 0 \\ 0 & 0 & 1 & -1 & 0 & 1 \end{array}\right)$$

$$\xrightarrow{\text{-3R3 + R1}} \left(\begin{array}{ccc|ccc} 1 & 0 & 0 & 7 & -3 & -3 \\ 0 & 1 & 0 & -1 & 1 & 0 \\ 0 & 0 & 1 & -1 & 0 & 1 \end{array}\right)$$

* This example is from "Purplemath.com," a great Web site for math students.

The left-hand side of the augmented matrix is the identity matrix $\mathbf{I}_3$. The right-hand side contains the inverse matrix of

$$\begin{pmatrix} \mathbf{1} & \mathbf{3} & \mathbf{3} \\ \mathbf{1} & \mathbf{4} & \mathbf{3} \\ \mathbf{1} & \mathbf{3} & \mathbf{4} \end{pmatrix}$$

Computations involving matrices are tedious to do by hand and it is easy to make mistakes. The techniques shown here are to illustrate the process. Modern calculators* can do all of this easily and accurately. However, you should understand the process.

Applications of matrices

Matrices are useful also for solving multivariate linear systems of equations** and to rotate graphs, among other applications.

Row echelon form of a square matrix

We first define a special type of matrix called the *row echelon form*; it is an upper triangular matrix in which the leading entry (i.e., the element in the diagonal) is 1 except in the last row, where it may be 0.

Solving systems of linear equations

First, set up the augmented *matrix of variable coefficients* and the constants on the right side of the equations:

Reduced Row Echelon Form ("rref").

A matrix is transformed into *reduced row echelon form* by using *elementary row operations.*

Example using *rref* to solve a system of three equations:

$x + 3y - 4z = 15$ (1)

$2x - y + 2z = -1$ (2)

$x + y - z = 6$ (3)

which has corresponding matrix with rows (from top to bottom) R1, R2, and R3

$$\left(\begin{array}{ccc|c} 1 & 3 & -4 & 15 \\ 2 & -1 & 2 & -1 \\ 1 & 1 & -1 & 6 \end{array}\right) \begin{array}{l} \text{R1} \\ \text{R2} \\ \text{R3} \end{array}$$

The columns, of course, represent the variables x, y, and z while the fourth column (separated by a vertical line) are the values on the right side of the " = " sign.

* Visit matrix.reshish.com.

** Linear equations with two or more variables with no exponents, e.g., $2x - 12y + 3z = 35$

The objective is to use elementary row operations to achieve the following matrix:

$$\left(\begin{array}{ccc|c} 1 & 0 & 0 & a \\ 0 & 1 & 0 & b \\ 0 & 0 & 1 & c \end{array}\right)$$

where the first three columns (and rows) are the identity matrix and the fourth column holds the solution: x, y, z. Converting the matrix yields the three equations $x = a$; $y = b$; and $z = c$. Note that R3 has a "1" in its second column, so we exchange (swap) R2 and R3 to get

$$\begin{array}{lr} x + 3y - 4z = 15 & (1) \\ x + y \; - z \;\; = 6 & (3) \\ 2x - y + 2z \; = -1 & (2) \end{array}$$

R2 ↔ R3 and for convenience, rename the rows as R1, R2, and R3. All we have done was to rearrange the order of equations to get a "1" in the second column in of the second row as follows:

$$\begin{array}{c} R1 \\ R2 \\ R3 \end{array} \left(\begin{array}{ccc|c} 1 & 3 & -4 & 15 \\ 1 & 1 & -1 & 6 \\ 2 & -1 & 2 & -1 \end{array}\right)$$

Next, we must change the "2" in bottom left to a "0." Let's multiply R2 by -2 and add the row to R3 to make a new R3: -2R2 + R3 → R3

$$\begin{array}{c} R1 \\ R2 \\ R3 \end{array} \left(\begin{array}{ccc|c} 1 & 3 & -4 & 15 \\ 1 & 1 & -1 & 6 \\ 0 & -3 & 4 & -13 \end{array}\right)$$

Now, add R3 to R1 to get two more zeros: R3 + R1 → R1

$$\begin{array}{c} R1 \\ R2 \\ R3 \end{array} \left(\begin{array}{ccc|c} 1 & 0 & 0 & 2 \\ 1 & 1 & -1 & 6 \\ 0 & -3 & 4 & -13 \end{array}\right)$$

(Multiply R1 by -1 and add it to R2: -1R1 + R2 → R2

$$\begin{array}{c} R1 \\ R2 \\ R3 \end{array} \left(\begin{array}{ccc|c} 1 & 0 & 0 & 2 \\ 0 & 1 & -1 & 4 \\ 0 & -3 & 4 & -13 \end{array}\right)$$

Multiply R2 by 3 and add it to R3: 3R2 + R3 →R3

$$\begin{array}{c} R1 \\ R2 \\ R3 \end{array} \left(\begin{array}{ccc|c} 1 & 0 & 0 & 2 \\ 0 & 1 & -1 & 4 \\ 0 & 0 & 1 & -1 \end{array}\right)$$

Note that by inspecting the matrix we see that $x = 2$ and $z = -1$. We could now go to the original equations and easily determine the value of y. However, let's continue with one more elementary row operation on the matrix.

Finally, add R3 to R2: R3 + R2 → R2

$$\begin{matrix} R1 \\ R2 \\ R3 \end{matrix} \left(\begin{array}{ccc|c} 1 & 0 & 0 & 2 \\ 0 & 1 & 0 & 3 \\ 0 & 0 & 1 & -1 \end{array}\right)$$

The solution is now direct: $x = 2$, $y = 3$ and $z = -1$ which is easily verified by "plugging in" the values obtained:

$$2 + 3\cdot3 - 4\cdot(-1) = 15 \qquad (1)$$
$$2 + 3 - (-1) = 6 \qquad (3)$$
$$2\cdot2 - 3 + 2\cdot(-1) = -1 \qquad (2)$$

Summary of elementary row operations

Elementary row operations on a matrix are simply allowable operations on systems of equations:

1. Add a multiple of any row to any other row.
2. Multiply any row non zero constant.
3. Exchange (swap) any pair of rows.

A matrix is in *Reduced Row Echelon Form* if

1. The first non-zero entry in any row is the integer "1" (called a *pivot*)
2. A "1" is the only non-zero entry in any column (except the last) or row.
3. The "1"s form a diagonal with the first entry in the top left.
4. Any rows of all zeros are at the bottom.

Graphing calculators such as the TI-83 or TI-84 have "rref" operators which will also do the job.

Cramer's Rule

An alternate method for solving systems of equations is through the use of *Cramer's Rule* which involves only determinants. Let's take the example of the system already solved by elementary row operations.

$$x + 3y - 4z = 15 \qquad (1)$$
$$2x - y + 2z = -1 \qquad (2)$$
$$x + y - z = 6 \qquad (3)$$

Let **D** be the determinant of the coefficient matrix

$$\mathbf{D} = \begin{vmatrix} 1 & 3 & -4 \\ 2 & -1 & 2 \\ 1 & 1 & -1 \end{vmatrix} = 1\cdot(1 - 2) - 2\cdot(-3 + 4) + 1\cdot(6 - 4) = -1$$

Let $\mathbf{D}_x$ be the determinant where the constants on the right side of the " = " of the equations replace the elements of the first column (the coefficients of x).

$$\mathbf{D}_x = \begin{vmatrix} 15 & 3 & -4 \\ -1 & -1 & 2 \\ 6 & 1 & -1 \end{vmatrix} \; 1 - 2) - (-1)\cdot(-3 + 4) + 6\cdot(6 - 4) = -2$$

Let $\mathbf{D}_y$ be the determinant where constants on the right side of the equations replace the elements of the second column (the coefficients of y).

$$\mathbf{D}_y = \begin{vmatrix} 1 & 15 & -4 \\ 2 & -1 & 2 \\ 1 & 6 & -1 \end{vmatrix} = 1\cdot(1 - 12) - 2\cdot(-15 + 24) + 1\cdot(30 - 4) = -3$$

Finally, let $\mathbf{D}_z$ be the determinant where constants on the right side of the equations replace the elements of the third column (the coefficients of z).

$$\mathbf{D}_z = \begin{vmatrix} 1 & 3 & 15 \\ 2 & -1 & -1 \\ 1 & 1 & 6 \end{vmatrix} = 1\cdot(-6 + 1) - 2\cdot(18 - 15) + 1\cdot(-3 + 15) = 1$$

Cramer's Rule tells us that the solution to this system of equations is

$$\left(\frac{\mathbf{D_x}}{\mathbf{D}}, \frac{\mathbf{D_y}}{\mathbf{D}}, \frac{\mathbf{D_z}}{\mathbf{D}}\right)$$

Substituting the values, the solution is

$$\frac{-2}{-1}, \frac{-3}{-1}, \frac{1}{-1}; \; (x, y, z) = (2, 3, -1)$$

which is the solution already found with the "rref" technique.

Other applications of matrices

There are many applications of matrices in mathematics, finance, network flows and graphs, etc. Modern calculators and computers can solve readily such problems. Do a Web search to find the many determinant calculators for checking your own work.

●

THE MATH TUTOR'S FRIEND
$\log_a X = \log_b X/\log_b a$
x^a
$\log_e(x) = \ln(x)$
e^x
a^x
$\log(XY) = \log(X) + \log(Y)$
$\left(\frac{x^2y^3z^{-4}}{y^{-1}}\right)^3$
It's easy !!
Mr. Math
$\ln(x)$
EXPONENTS & LOGARITHMS

6. EXPONENTS & LOGARITHMS
TABLE OF CONTENTS

•

6. EXPONENTS & LOGARITHMS

Exponents

Informally, an *exponent* (or *power*) is a number that specifies the number of times a number, variable, term, or expression is multiplied by itself; thus, $4^3 = 4 \times 4 \times 4 = 64$; $(x^2 + 1)^2 = (x^2 + 1) \times (x^2 + 1)$; etc. The number related to an exponent, is the *base*. See Figure 6-1.

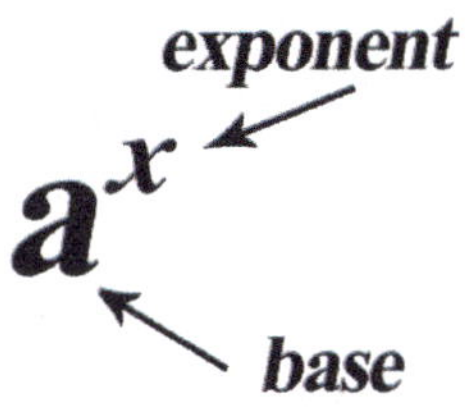

Figure 6-1

The three functions $f(x) = 2^x$, $f(x) = 4^x$, and $f(x) = 10^x$ shown in Graph 6-1 all have y-intercepts at $y = 1$. This is because $x = 0$ and anything raised to a zero power equals 1. Note also that there is a horizontal asymptote ($y = 0$) and as the

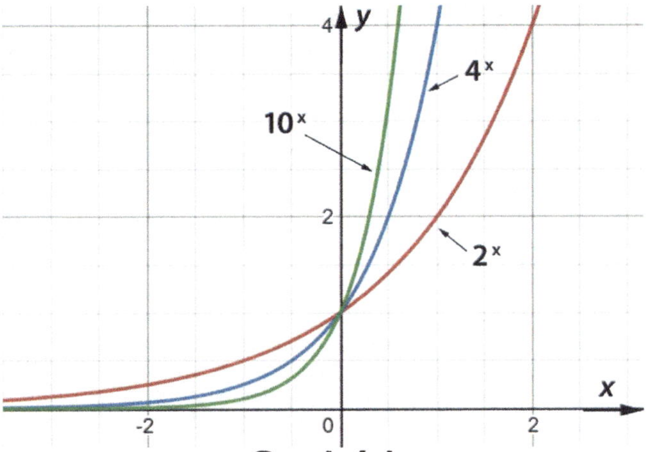

Graph 6-1

value of the base increases, the function becomes "steeper."

Rules for exponents

Product rule. There are three possibilities: an algebraic term may contain

(1) identical *bases* (e.g., $b^2 b^4 b^x$);
(2) different bases & different exponents, (e.g., $b^2 a^3 c^x$); and
(3) different bases and a common *exponent* (e.g., $b^2 a^2 c^2$).

<u>Examples</u>

(1)	$x^a x^b x^c = x^{(a+b+c)}$	(simply add the exponents)
(2)	$b^2 a^3 c^x = b^2 a^3 c^x$	(different bases & different exponents; can't change anything)
(3)	$b^2 a^2 c^2 = (bac)^2$	(product of the bases with the same exponent)

This rule also works "backwards," that is

$$x^{(a+b+c)=} = x^a x^b x^c \text{ and } (bac)^2 = b^2 a^2 c^2$$

Example of product rule (make sure you understand this):

$$(x^r x^s x^t b^2 a^3 c^y b^2 a^2 c^2) = x^{(r+s+t)} b^4 a^5 c^{(y+2)}$$

Quotient rule. There are three possibilities: an algebraic term may contain

(1) identical bases (e.g., b^2/b^4);
(2) different bases & different exponents, (e.g., b^2/a^3); and
(3) different bases and a common exponent (e.g., $b^2/(a^2c^2)$).

Examples:

(1) $x^a/x^b = x^{(a - b)}$ (subtract the denominator exponent from that of the numerator)

(2) $b^2/a^3 = b^2/a^3$ (different bases & exponents; can't change anything)

(3) $b^2/(a^2c^2) = (b/ac)^2$ (different bases, same exponent)

As with the product rule, the quotient rule works "backwards."

Example demonstrating both product and quotient rule:

$$\frac{x^2y^3a^xb^2c^2}{xy^4a^2d^2} = \frac{x}{y}a^{(x-2)}\left(\frac{bc}{d}\right)^2 = xy^{-1}a^{(x-2)}(bcd^{-1})^2$$

Yes, you can have negative exponents; we'll discuss them shortly.

Powers of exponents rules

Example of powers rule

$$(3^2)^3 = (3{\cdot}3)^3 = (9)^3 = 729 = 3^6 = 729$$

$$2^{4^2} = 2^{(4^2)} = 2^{16} = 65,536$$

$$(x^a)^b = x^{ab}$$

$$x^{a^b} x^{(a^b)}$$

Exponents and radicals rule

$$x^{\frac{n}{m}} = \sqrt[m]{x^n}$$

Examples of radicals rule

$$x^{\frac{1}{2}} = \sqrt[2]{x} = \sqrt{x}$$ (the radical index 2 is implied)

$$x^{\frac{3}{4}} = \sqrt[4]{x^3}$$ (the numerator of a fractional exponent is the power of x; the denominator is the radical index)

Negative exponent rule

Moving a base with an exponent from numerator to denominator or vice-versa merely changes the sign of its exponent; that is, a negative exponent becomes positive and vice-versa.

$$x^{-a} = \frac{1}{x^a} \qquad \frac{1}{x^{-b}} = x^b$$

Examples

$$3^{-1} = 1/3; \; 1/(2^{-2}) = 4; \frac{a^{-2}}{b^{-3}} = \frac{b^3}{a^2}$$

Special cases of exponents rule

Zero power and zero base rules

$$n^0 = 1 \text{ for } n \neq 0$$

$$0^n = 0; n > 0.$$

0^0 may be 1, but is better treated as undefined

Why the restriction $n > 0$? Well, let's try 0^{-n}; that is $1/(0^n)$ and we know that division by 0 is not allowed because it is undefined.

Unity power and unity base rules

$$b^1 = b$$

$$1^n = 1$$

Negative unity rule

$$(-1)^n = 1 \text{ if } n \text{ is even or zero;}$$

$$(-1)^n = -1 \text{ if } n \text{ is odd.}$$

This, of course, is an application of the rules for signed numbers.

How to handle problems involving exponents

Frequently, the student is asked to simplify complicated-looking expressions involving factors with a variety of exponents. Here are two examples with tips for solutions:

$$\sqrt{8x^3y^4z^7}$$

This is a square root monomial with four factors. When the radical index is *even*, i.e., 2, 4, 6, *etc*, then find each factor in the radicand with a power divisible by the index (in this example we look for powers divisible by 2). Factors with odd powers can be expressed as the product of factors, one of which has an even power, by applying Rule 1 of the product rule. Thus, 8 can be expressed as $2{\cdot}2^2$; x^3 as $x{\cdot}x^2$; and z^7 as $z{\cdot}z^6$, so that each original factor is the product of two factors, one of which is divisible by 2. Therefore, the example problem becomes

$$\sqrt{2 \cdot 2^2 x \cdot x^2 y^4 z \cdot z^6}$$

Rewrite using the *exponents and radicals rule* to remove the radical:

$$(2 \cdot 2^2 \cdot x \cdot x^2 \cdot y^4 \cdot z \cdot z^6)^{\frac{1}{2}}$$

Applying the powers of exponents rule, the above becomes

$$2^{1/2}2x^{1/2}xy^2z^{1/2}z^3$$

Now, group factors with the same exponents:

$$2^{1/2}x^{1/2}z^{1/2}2xy^2z^3 = (2xz)^{1/2}2xy^2z^3$$

with result

$$\sqrt{8x^3y^4z^7} = 2xy^2z^3\sqrt{2xz}$$

<u>Another example involving radicals and exponents</u>

Simplify

$$\sqrt[3]{\frac{27x^2y^{-4}z^8}{x^{-1}y^3z}}$$

First, combine *like* factors (the xs, ys and zs) to get positive exponents using the *negative exponents* and *product/quotient of exponents* rules:

$$\sqrt[3]{\frac{27x^3z^7}{y^7}}$$

The radical index is 3 (cube root), so we want to "extract" factors that have exponents divisible by 3 after manipulation using the product rule for exponents. Thus $27 = 3^3$; $z^7 = zz^6$; and $y^7 = yy^6$ and so the problem becomes

$$\sqrt[3]{\frac{3^3x^3zz^6}{yy^6}}$$

"Extracting" factors whose exponents are divisible by 3 from the radical, we have

$$\frac{3xz^2}{y^2}\sqrt[3]{\frac{z}{y}}$$

Both of these examples illustrate the procedure of looking for exponents divisible by the radical index (do this after grouping like factors) and is really the "trick" for handling complicated radical expressions with exponents.

Logarithms

Briefly, the logarithm of a number is the power, *a*, to which a fixed number, *b*, called the base must be raised in order to produce a given number, *c*, (the argument of the logarithm). This is illustrated below:

$$\log_b(c) = a$$

"b" is the base

$$b^a = c$$

In logarithm notation, when no base is shown, assume a base of 10; thus "log(x)" is the same as "$\log_{10}(x)$".

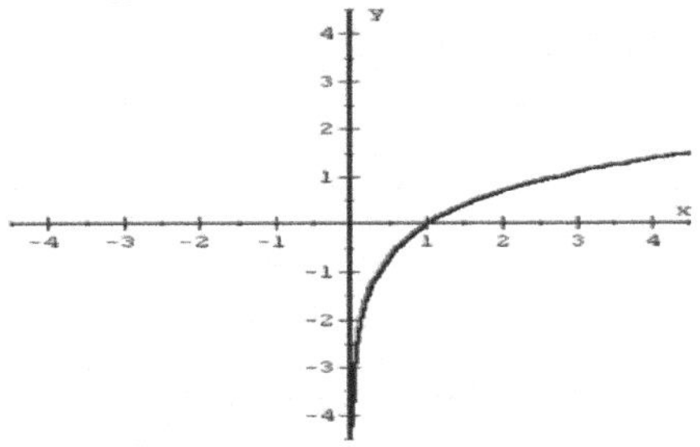

$f(x) = \log_a x$

Logarithmic

Rules for logarithms

Logarithms, as do exponents, change multiplication and division to addition and subtraction, respectively.

1) $\log(xy) = \log(x) + \log(y)$ [compare to $b^x b^y = b^{(x+y)}$]
2) $\log(x^a) = a \cdot \log(x)$ [compare to $(x^a)^b = x^{ab}$]
   This is an application of the product rule:
3) $\log(x/y) = \log(x) - \log(y)$ [compare to $b^x/b^y = b^{(x-y)}$]
   This is an application of the exponent rule where $a = -1$
   $$\log(x/y) = \log(x \cdot y^{-1}) = \log(x) + (-1)\log(y) = \log(x) - \log(y)$$

Don't forget the special cases:

$\log_b b = 1$ because $b^1 = b$; and $\log_b 1 = 0$ because $b^0 = 1$.

Natural Logarithms

There is a number, ***e***, called Euler's constant which is approximately 2.7182818. Both irrational and transcendental, ***e*** is commonly used as a base for logs and as a base for exponents:

$\log_e(x) = \ln(x)$ is called the "natural log of x."
["loge(x)" is sometimes used instead of $\log_e(x)$]

The rules of logarithms also apply to $\ln(x)$, and so

$$\ln(e^x) = x \ln(e) = x \times 1 = x \text{ because } \ln(e) = \log_e e = 1$$

Change of base formula

Suppose you want to find the logarithm of a given base using a logarithm of different base. For example, most calculators have only a <*log*> button and a <*ln*> button, so if you need to evaluate the logarithm of a number expressed in a different base you must employ the easy-to-remember *change-of-base formula*:

$$\log_b(c) = \frac{\log_d(c)}{\log_d(b)}$$

Evaluating $\log_b(c)$ using a logarithm of base d

Simply take the log of the argument, *c*, in the new base, *d*, and divide it by the logarithm of the old base, *b*, in the new base, *d*, as shown.

<u>Example</u>: $\log_2(15) = \log(15)/\log(2) \approx 3.91$ This allows you to find the log to the base 2 of 15 using a standard scientific or graphic calculator. Try it! It doesn't matter whether you use ***log*** or ***ln***, you'll get the same result:

$$\log_2(15) = \ln(15)/\ln(2) \approx 3.91.$$

Simplifying fractions having radicals in the denominator

When an algebraic problem is in its final form, it is preferable to not have any fractions with radicals in the denominator. Examples of these are rational functions (i.e., fractions) with denominators that include square root, cube roots, and any parenthetical expression having a fractional exponent.

<u>Examples:</u> $1/\sqrt{2}$; $(x^2 - 3x + 2)/(x + 5)^{1/2}$; $\frac{7}{\sqrt[3]{6}}$ and $\frac{x-3}{x+\sqrt{5}}$

Rationalize the denominator

How do we "remove" the radical expressions from the denominator? The solution is to *rationalize the denominator*. For a single radical term (the 1st three examples above), just multiply both numerator and denominator by the denominator:

$$\frac{1}{\sqrt{2}} = \frac{1}{\sqrt{2}} \cdot \frac{\sqrt{2}}{\sqrt{2}} = \frac{\sqrt{2}}{2}$$

Next,

$$(x^2 - 3x + 2)/(x + 5)^{1/2}$$
$$= [(x^2 - 3x + 2)/(x + 5)^{1/2}] \cdot [(x + 5)^{1/2}/(x + 5)^{1/2}]$$
$$= (x^2 - 3x + 2)\,(x + 5)^{1/2}/\,(x + 5)$$

In the first two cases above we used the "trick" of multiplying the term by the fraction

$$\frac{\text{original denominator}}{\text{original denominator}}$$

which, of course, equals 1, and multiplying by 1 changes nothing.

The denominator in the third example is $\sqrt[3]{6}$ which is also $6^{1/3}$. Note that

$$6^{\frac{1}{3}} \cdot 6^{\frac{2}{3}} = 6^{\frac{3}{3}} = 6^1 = 6$$

adding 2/3 to the exponent will make it equal to 1. This is done using the law of exponents, so we multiply the expression by a fraction with $\sqrt[3]{6}$ as both numerator and denominator; in other words, we multiply the expression by 1 and the denominator no longer contains a radical.

$$\frac{7}{\sqrt[3]{6}} = \frac{7}{\sqrt[3]{6}} \cdot \frac{\sqrt[3]{6^2}}{\sqrt[3]{6^2}} = \frac{7\sqrt[3]{6^2}}{\sqrt[3]{6^3}} = \frac{7\sqrt[3]{36}}{6}$$

Before discussing the last example, $\frac{x-3}{x+\sqrt{5}}$ we need to introduce the *conjugate*.

Given a binomial $(a + b)$, its conjugate is $(a - b)$. It's as simple as that! Also, recall the pattern of *difference of squares*:

$$(a + b)(a - b) = a^2 - b^2.$$

Therefore, if we multiply the denominator of the last example by its conjugate we will get the difference of squares, which in this case is $x^2 - 5$, as shown below.

$$\frac{x-3}{x+\sqrt{5}} = \frac{x-3}{x+\sqrt{5}} \cdot \frac{x-\sqrt{5}}{x-\sqrt{5}} = \frac{(x-3)(x-\sqrt{5})}{x^2-5}$$

In these examples, the radical in the denominator was removed by the technique shown. There are cases in which rationalizing the denominator is not necessary. For example, if the expression is an intermediate result with more computation to follow, or if the expression is to be evaluated for a specified value for x.

Exponential functions

The *exponential function* f(x) with base $a > 0$ and $a \neq 1$ is denoted by a^x, where a and x are real numbers. The figure to the right shows the exponential function $f(x) = 2^x$.

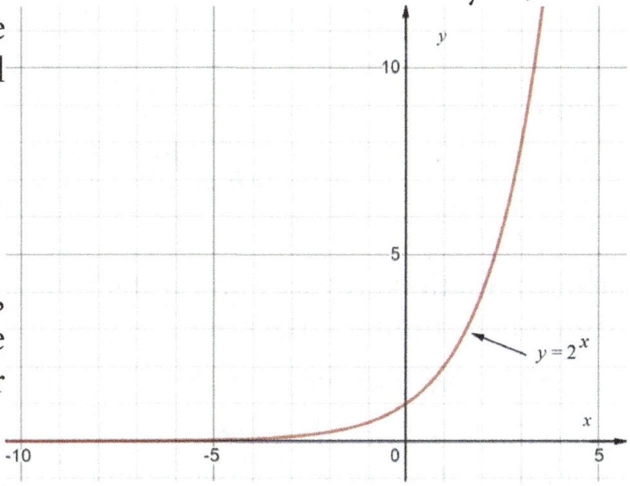

Exponential growth

Consider the function

$$f(t) = a \cdot b^{t/\tau}$$

where t is time, a is the *initial value*, $b > 0$ is the *growth factor*, τ is the *time constant* (the time required for x to increase by one factor of b).

<u>Example</u>:

A population with initial size $a = 100$ doubles (grows by a factor of 2; i.e., $b = 2$) every 5 minutes (i.e., $\tau = 5$). The population after one hour ($t = 60$ minutes) is

$$f(60) = a \cdot b^{t/\tau} = 100 \cdot 2^{60/5} = 100 \cdot 2^{12} = 100 \cdot 4{,}096 = 409{,}600.$$

Another exponential formula is

$$P(t) = P_0 e^{kt} \qquad (6\text{-}1)$$

where

$P(t)$ is the population at time t

P_0 is the initial population

t is the specified elapsed time

k is a constant to be determined

For the example we have $P_0 = 100$ and $P(5) = 200$ (doubling every 5 minutes) then applying Equation (6-1) we have

$$200 = 100\boldsymbol{e}^{k5} \text{ or } 2 = \boldsymbol{e}^{k5}$$

Take the natural logarithm of both sides:

$$\ln(2) = \ln(\boldsymbol{e}^{k5}) = \ln(\boldsymbol{e}^{k})^5 = 5\ln(\boldsymbol{e}^{k}) = 5k \qquad [\text{because } \ln(\boldsymbol{e}^{k}) = k]$$

from which $k = \ln(2)/5$

In the example we want the population at $t = 60$ minutes, so from Equation (1)

$$P(60) = e^{\frac{\ln(2)}{5}\cdot 60} = 100e^{\ln(2)\cdot 12} = 100(e^{\ln(2)})^{12} = 100 \cdot 2^{12} = 409{,}600$$

Note that $e^{\ln 2} = 2$

This is the same result as before; that means that the two expressions are equal:

$$ab^{\frac{t}{\tau}} = ae^{kt}$$

Divide through by a and take the natural logarithm of both sides,

$$\frac{t}{\tau}\ln b = kt$$

From which

$$k = \frac{\ln b}{\tau} = \frac{\ln[growth\ factor]}{[time\ constant]}$$

The population grows by amount b (the growth factor) every τ (time constant) time units (minutes, days, etc.).

Another formula for growth

Take the function $f(t) = a{\cdot}b^{t/\tau}$. The base, b, is the growth factor, that is, growth at a rate r, the percent increase in period τ. In the example $r = 100\%$; that is $b = (1 + r/100)$, so the above function becomes $f(t) = a{\cdot}(1 + r/100)^{t/\tau}$

Summary of the forms for exponential growth

Suppose $b = \boldsymbol{e}$. then the following are equivalent forms for exponential growth:

$$f(t) = a{\cdot}b^{\,t/\tau} = P_0\boldsymbol{e}^{\,kt} = a{\cdot}(1 + r/100)^{\,t/p}$$

where

a is the initial value, P_0 or f(0), that is, the amount at $t = 0$;
b and $(1 + r/100)$ are each a factor of increase;
r is the percent increase in period τ;
τ is the time constant (time required for t to increase by one multiple of b).

In the special case of doubling a population every τ time units (the "*doubling time*"), we have $b = 2$ and $f(t) = a{\cdot}2^{\,t/\tau}$

Logistic function

Parent logistic function

The *parent (basic) logistic function* is

$$f(x) = \frac{1}{1 + e^{-x}}$$

The basic logistic function appears at the bottom in Graph 6-2 below. The value of the parent logistic function ranges from 0 to 1. Just as with any other function we can translate the logistic function left or right and up or down.

$$\frac{L_1}{1 + e^{-k(x - x_0)}} + L_2$$

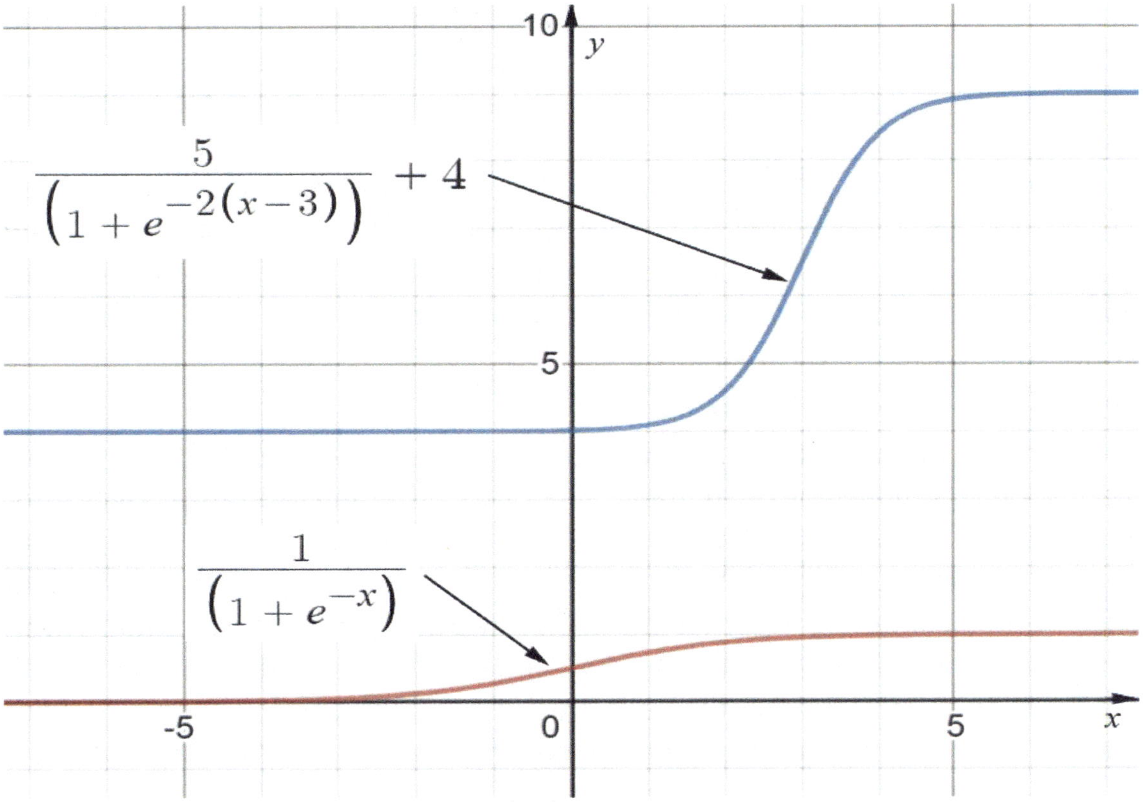

Graph 6-2

The logistics function is an S-shaped *(sigmoid) curve* with equation where

L_1 is the maximum value, or *carrying capacity*;
L_2 moves the curve along the y-axis by the amount L_2;
k is the *growth rate* (*steepness*); and
x_0 the x-value of the sigmoid's midpoint (the distance the function moves along the x-axis). See the diagram below: $k = 2$, $L_1 = 5$; $x_0 = 3$, and $L_2 = 4$.

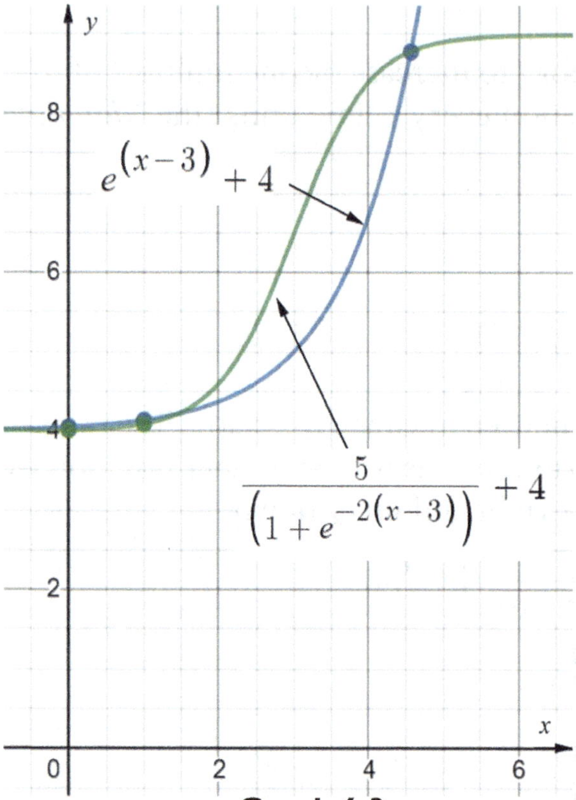

Graph 6-3

Also called the *logistic curve*, its domain is all real numbers (-∞, ∞). The logistic function has important applications in economics, ecology, statistics, and many other fields. Exponential growth and the logistic function have shapes, roughly "J" and "S," respectively. The former is capped and the latter ranges from 1 to ∞. See the Graph 6-3 on the left for a comparison of the *J-shaped exponential* and *S-shaped (sigmoid) logistic functions.*

Periodic compounding

Nominal interest rate

In finance and economics, the *nominal interest rate* or *nomial rate of interest* can be either the rate of interest before adjustment for inflation or the interest rates "as stated" without adjustment for the full effect of compounding (also referred to as the *nominal annual rate*). An interest rate is called nominal if the frequency of compounding (e.g. a month) is not identical to the basic time unit in which the nominal rate is quoted (normally a year).

Nominal versus real interest rate

The nominal interest rate, or *Annualized Percentage Rate* (*APR*), is the periodic interest rate multiplied by the number of periods per year. For example, a nominal annual interest rate of 12% based on monthly compounding is a 1% interest rate per month. Do not compare APRs unless the compounding periods are the same; effective interest rates correct for this by "converting" nominal rates into annual compound interest. Some interest rates quoted in advertisements, for example, are based on nominal, rather than effective, interest rates, and may understate the interest rate compared to the equivalent effective annual rate. APR should not be confused with *simple interest,* which is not compounded.

Effective interest rate

Calculate *effective interest rate* as if compounded annually: let r be the effective rate, *i* the nominal rate as a decimal, e.g., 8% = 0.08), and *n* the number of compounding periods per year (for example, $n = 12$ for monthly compounding):

Monthly compounding

Example: A nominal interest rate of 6% compounded monthly is equivalent to an effective interest rate of 5.12%.

$r = (1 + 0.05/12)^{12} - 1 = 1.0512 - 1 = 0.0512$ (rounded to four decimal places), or 5.12%

Daily compounding

Example: A loan with a 9% nominal annual rate and daily compounding has an effective annual rate of 9.38%. For a loan of $10,000 (paid at the end of the year in a single lump sum), the borrower would pay $38.07 more than one who was charged 9% interest, compounded annually (for one year).

Accumulated value

The total accumulated value (the principal plus compounded interest), is given by

$$A = P(1 + r/n)^{nt}.$$

Where

P is the original principal sum | A is the new principal sum

r is the nominal annual interest rate | n is the compounding frequency

t is the overall length of time the interest is applied (usually years).

Accumulation function

The principal P, is a coefficient and may be dropped for simplicity, resulting in the

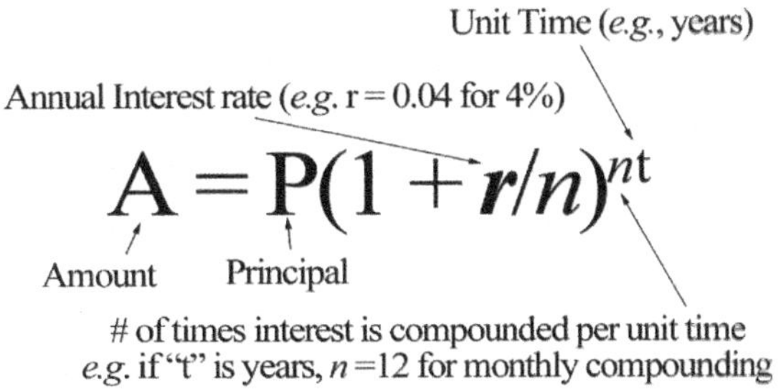

accumulation function, $\boldsymbol{a}$(t). The accumulation function shows the value of $1 over any length of time. The accumulation function for simple and compound interest is

$$\boldsymbol{a}(t) = (1 + r/n)^{nt}.$$

Simple interest is the accumulation function with $n = 1$.

Continuous compounding

Continuous compounding is achieved when *n*, the number of compounding periods per year, increases without limit, and can be thought of as letting *n* approach infinity (i.e., $n \to \infty$). The amount after "t" periods of continuous compounding can be expressed as $A = Pe^{rt}$.

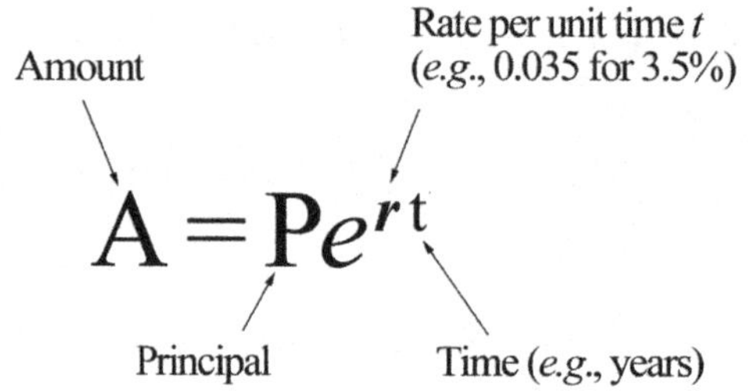

THE MATH TUTOR'S FRIEND
Fibonacci
progressions
Sigma notation
geometric series
Infinite Series
Domain
It's easy !!
Mr. Math
SEQUENCES & SERIES

7. SEQUENCES & SERIES
TABLE OF CONTENTS

•

7. SEQUENCES AND SERIES

Introduction

A *sequence* is a set of numbers written in some particular order for example, 1, 3, 5, 7, 9, ... The set of three dots (the 4th dot is a period) following the last comma is an *ellipsis* and indicates that the sequence continues on indefinitely; however, if an ellipsis is itself followed by a comma and a variable, say *n*, then the sequence is *finite* and ends with the number "*n*." Clearly, the example is a sequence of odd numbers. The first number is 1 (an odd number); each successive number is obtained by adding 2 to the preceding number. This sequence is an *arithmetic** *sequence* because there is a common difference ("2") between adjacent numbers in the sequence.

HINT: Take any integer and call it "*n*,"which may be odd or even. To get an even number then write "2*n*;" for an odd number, write $2n + 1$ (or $2n - 1$).

Some sequences conform to a rule, such as the sequence of square numbers (the *n*th term is n^2), cubes, the sequence of prime numbers, etc.

Sequences, series, and progressions

As stated above, a *sequence* is a set of numbers arranged according to some rule; a *series* is a sequence in which the terms are separated by " + " signs, and a *progression* is a sequence that can be defined by a specific mathematical formula like "the *n*th term is $(2n + 1)$." All series are based on specific sequences, and although all progressions are sequences, all sequences are not progressions (for example, the sequence of prime numbers – a formula for determining the *n*th prime number, if there is one, is yet to be discovered). Here's a sequence that follows a different rule:

$$1, -1, 1, -1, 1, -1, \ldots$$

is an *infinite sequence* of numbers alternating between 1 and −1. For example, a *finite sequence* such as 3, 6, 17, 101, which consists of four positive integers in ascending order, is an *ordered set* (3, 6, 17, 101); that is, the members of the set are in a fixed order. If we are talking simply about the group of numbers where order is unimportant then we have the *unordered set* {3, 6, 17, 101} which is the same as the set {17, 3, 101, 6}. Braces ("{ }") depict a set of no particular ordering. The numbers 1, 3, 5, 7, 9, . . ., $2n + 1$ form a finite sequence of odd integers. This is a progression because the *n*th term is computed with the formula. Since there are an infinite number of odd numbers, the sequence can be expressed as 1, 3, 5, 7, 9, ..., $(2n + 1)$,... . It is convenient to express a sequence as an arrangement of *subscripted* literal numbers such as a_i, u_i, etc. Progressions, which are functions, can be expressed with a formula, $f(n)$.

* "Arithmetic" here is an adjective pronounced with the accent on the penultimate (next-to-the-last) syllable.

Other Types of Sequences

Fibonacci Sequence (O.E.I.S. A000045*)

$$0, 1, 1, 2, 3, 5, 8, 13, 21, 34, 55, 89, 144, 233, \ldots, f(n), \ldots.$$

is a progression with formula $f(n) = f(n-1) + f(n-2)$ with $f(0) = 0$ and $f(1) = 1$.

Series

A *series* is formed from a sequence simply by adding all the terms together. For example, suppose we have the sequence $u_1, u_2, u_3, \ldots, u_n$. The series we obtain from this is $u_1 + u_2 + u_3 + \ldots + u_n$, and we write S_n for the sum of these n terms.

<u>Example</u>:
Consider the sequence of numbers

$$1, 2, 3, 4, 5, 6, \ldots, n.$$

The sum $S_1 = 1$ is simply the first term on its own. The sum of the first two terms is $S_2 = 1 + 2 = 3$. Continuing, $S_3 = 1 + 2 + 3 = 6$ and $S_4 = 1 + 2 + 3 + 4 = 10$.

Arithmetic sequence

Consider the sequence

$$1, 4, 7, 10, \ldots. \qquad \text{(Sequence 7-1)}$$

Sequence (7-1) is formed by starting with a first term and then generating successive terms by adding 3 to the previous term. That is, the *common difference* between consecutive terms in the sequence is the constant 3. We could also subtract a constant instead because that is just the same as adding a negative constant.** For example, in the sequence

$$10, 7, 4, 1, -2, -5, \ldots. \qquad \text{(Sequence 7-2)}$$

the difference between consecutive terms is -3. An *arithmetic sequence* is one in which a successive term is obtained by adding a constant to the previous term.

Let a be the first term of the sequence and let d stand for the common difference between successive terms. In general the *arithmetic series* can be written as

$$a_1, a_1 + d, a_1 + 2d, a_1 + 3d, a_1 + 4d, \ldots.$$

where $a_1 = 1$ is the first term, and $d = 3$ is the common difference. If we wanted to write down the nth term, we would have $a_1 + (n - 1)\,d$. This should be clear because if there are n terms there must be $(n - 1)$ common differences between successive terms, so we simply add $(n - 1)d$ to a_1, the first term in the sequence. [This is easily seen from this example: the sequence "1, 4, 7, 10" has $n = 4$ terms with a constant value of 3 between them and $(n - 1)$ "spaces," so the 4th term is $1 + (4 - 1)\times 3 = 10$ is verified.] In short, let a_n be the last term in the (finite) sequence. Then $a_n = a_1 + (n - 1)\,d$.

* The on-line encyclopedia of integer sequences (O.E.I.S), founded in 1964 by N. J. A. Sloane, contains thousands of different integer sequences and is certainly worth a visit. A number such as A020513 identifies each sequence and is supported by discussion, history, related theorems (and proofs), and is cross-referenced with citations and links.

** This, of course, includes minus signs because $a - b = a + (-b)$.

Geometric sequence

Consider the sequence

2, 6, 18, 54, . . . (Sequence 7-3)

Here, each term in the sequence is 3 times the previous term. And in the sequence 1, -2, 4, -8, . . . , each term is -2 times the previous term. Such a sequence is called a *geometric sequence* (or *geometric progression* or GP).

Let a be the first term, and note that rather than a common difference there is a common ratio; that is, the ratio of successive terms is always constant. Let r be the common ratio. The general geometric progression can be expressed as

$$a,\ ar,\ ar^{2},\ ar^{3}, \ldots,\ ar^{n}$$

The nth term of a geometric progression is ar^{n-1}, where the power $(n-1)$ is always one less than the position n of the term in the sequence. In Sequence (7-3), $a = 2$ and $r = 3$.

Sigma notation to express a sum

There is a "shortcut" to express of sums of the form $u_1 + u_2 + u_3 + \ldots + u_n$ and that is to use the capital form of the Greek letter *sigma*: $\sum$, which one may compare to the capital letter "S" (for *sum*).* Thus, the sum $a_0 + a_1 + a_2 + a_3 + \ldots + a_n$ is expressed as $\sum_{i=0}^{n} a_i$, where i, the *index of summation*, ranges from 0 to n.**

Sum of an arithmetic series

The sum of the first n terms of an arithmetic series is given by

$$S_n = n(a_1 + a_n)/2.$$

For example, the sum of the first 100 integers (1, 2, 3, ..., 100) is

$$S_{100} = 100(1 + 100)/2 = 50(1 + 100) = 50 + 5{,}000 = 5{,}050.$$

So, why doesn't the common difference, d, appear in the formula for the sum of an arithmetic series? One answer is that d is embedded in a_n by definition.***

Domain and range of a sequence

A sequence meets the conditions for a function; that is, there is a set of numbers, $i = 0, 1, 2, 3, \ldots, n$ which forms the *domain of the sequence* (function); the n following the dots tells us that this is a finite sequence. Each member of the domain has exactly one associated value, $(a_0, a_1, a_2, a_3, \ldots, a_i, \ldots, a_n)$. The *range of a sequence* is the difference between the largest and smallest number in the sequence. These are the numbers we use for counting, and we have included n of them. As before, the dots indicate that we have not written all the numbers down explicitly, and that the last number is a_n. Thus we may call a sequence $f(i)$ and so $f(4) = a_4$, for example. The sequence $f(n)$ can also be defined by a rule or a formula. For example, $f(n) = 2n - 1$.

* The lower-case sigma, σ, looks nothing like the capital sigma; one of its uses is to express *standard deviation* in statistics.

** Here, the letter i is not the imaginary number $i = \sqrt{-1}$. It is an index (in this case an integer between 0 and n).

*** Note that $a_n = a_0 + (n-1)$d which embeds the common difference, d.

Sum of a geometric series

Let S_n be the sum of the first n terms of a geometric sequence:

$$S_n = a + ar + ar^2 + ar^3 + \ldots + ar^{n-1}.$$

This is called a *geometric series*. The trick here to find the sum is to multiply S_n by r:

$$rS_n = ar + ar^2 + ar^3 + \ldots + ar^{n-1} + ar^n$$

now subtract rS_n from S_n:

$$S_n = a + ar + ar^2 + ar^3 + \ldots + ar^{n-1}$$
$$rS_n = \quad ar + ar^2 + ar^3 + \ldots + ar^{n-1} + ar^n$$
$$S_n - rS_n = a - ar^n$$

Factor S_n from the left side of the equation and factor a from the right side.

$$S_n(1 - r) = a(1 - r^n)$$

Now divide by $(1 - r)$, where $r \neq 1$, to give

$$S_n = a(1 - r^n)/(1 - r)$$

Where a is the common ratio and S_n is the sum of the first n terms of the series.

Infinite series

We have seen that the sum of a series may be expressed as S_n for the sum of the first n terms of that series, where n ranges anywhere from 1 to ∞. S_n is the *partial sum*, i.e., the sum of the first n terms of the series. An infinite series is the sum, S, of the partial sums:

$$\Sigma S_i = S_1 + S_2 + S_3 + S_4 + \ldots + S_\infty$$

Convergence and divergence of infinite series

If $\Sigma S_i = C$ where $-\infty < C < \infty$ then the infinite series is convergent to C; otherwise, the infinite series is divergent.

Note that the series for the sequence

$$1, -1, 1, -1, 1, -1, \ldots,$$

is divergent because it alternates and so does not converge to a known value. It is a geometric sequence with common ratio -1.

Sum of an infinite series with common ratio | r | < 1.

An *infinite geometric series* is the sum of an infinite geometric sequence. The general form of the infinite geometric series is $a + ar + ar^2 + ar^3 + \ldots$, where a is the first term and r is the common ratio. If $|r| \geq 1$ then the series diverges (gets larger and larger without bound). However, if $|r| < 1$, the series will converge to some finite number. The sum, S, of such an infinite geometric series is

$$S = a/(1 - r), \text{ where } |r| < 1.$$

Example: Find $\sum_{i=1}^{\infty} \frac{1}{2^n}$

Here, $r = ½$ (which is less than 1) and $a = 1$. Using the above formula we have

$$S = \frac{1}{1-1/2} = 2.$$

Partial sum

If the sequence of partial sums is a *convergent sequence* (i.e., if its limit exists and is finite) then the series is convergent also. Likewise, if the sequence of partial sums is a *divergent sequence* (i.e., its limit doesn't exist or is +∞ or -∞) then the series is also divergent.

The sum of infinite terms is an infinite series. Partial sums are sometimes called *finite series*. The *n*th partial sum is the sum of the first *n* terms of an infinite series.

Example of a converging series:

$$\sum_{n=1}^{\infty} \frac{1}{n(n+1)} = \frac{1}{n} - \frac{1}{n+1} = \left(\frac{1}{1} - \frac{1}{2}\right) + \left(\frac{1}{2} - \frac{1}{3}\right) + \left(\frac{1}{3} - \frac{1}{4}\right) + \ldots + \left(\frac{1}{n} - \frac{1}{n+1}\right)$$

Notice that the terms cancel out in pairs, leaving only the first and last terms (this is called a *telescoping series*). The partial sums, $S_1, S_2, \ldots, S_n$ are $S_1 = ½$; $S_2 = 2/3$; and so on. The terms of the series get closer and closer to 1 so that the series has limit 1.

$$\sum_{n=1}^{\infty} \frac{1}{n(n+1)} = \lim_{n\to\infty} \left[1 - \frac{1}{n+1}\right] = 1$$

Looks can be deceiving! Here are two series:

$$\sum_{i=1}^{\infty} \frac{1}{i} \text{ and } \sum_{i=1}^{\infty} \frac{1}{i^2}.$$

The first is divergent and the second converges to $\frac{\pi^2}{6}$.

Harmonic series

$$\frac{(n+1)}{n(n+1)} = \frac{A(n+1)}{n} + \frac{B(n+1)}{(n+1)}$$

$$\frac{1}{n} = \frac{A(n+1)}{n} + B$$

A special series, called the *harmonic series*, is given by

$$\frac{1}{1} + \frac{1}{2} + \frac{1}{3} + \frac{1}{4} + \frac{1}{5} + \frac{1}{6} + \frac{1}{7} + \ldots + \frac{1}{n},$$

a sum of the positive *vulgar fractions* that are *unit fractions* and can be shown to be a divergent series.

Formulas for special series

$$\sum_{i=1}^{n} 1 = n$$

$$\sum_{i=1}^{n} i = \frac{n(n+1)}{2}$$

$$\sum_{i=1}^{n} i^2 = \frac{n(n+1)(2n+1)}{6}$$

•

INDEX

Symbols

A

B

C

D

J

L

M

N

O

P

T

U

V

X

Y

Z

• • •

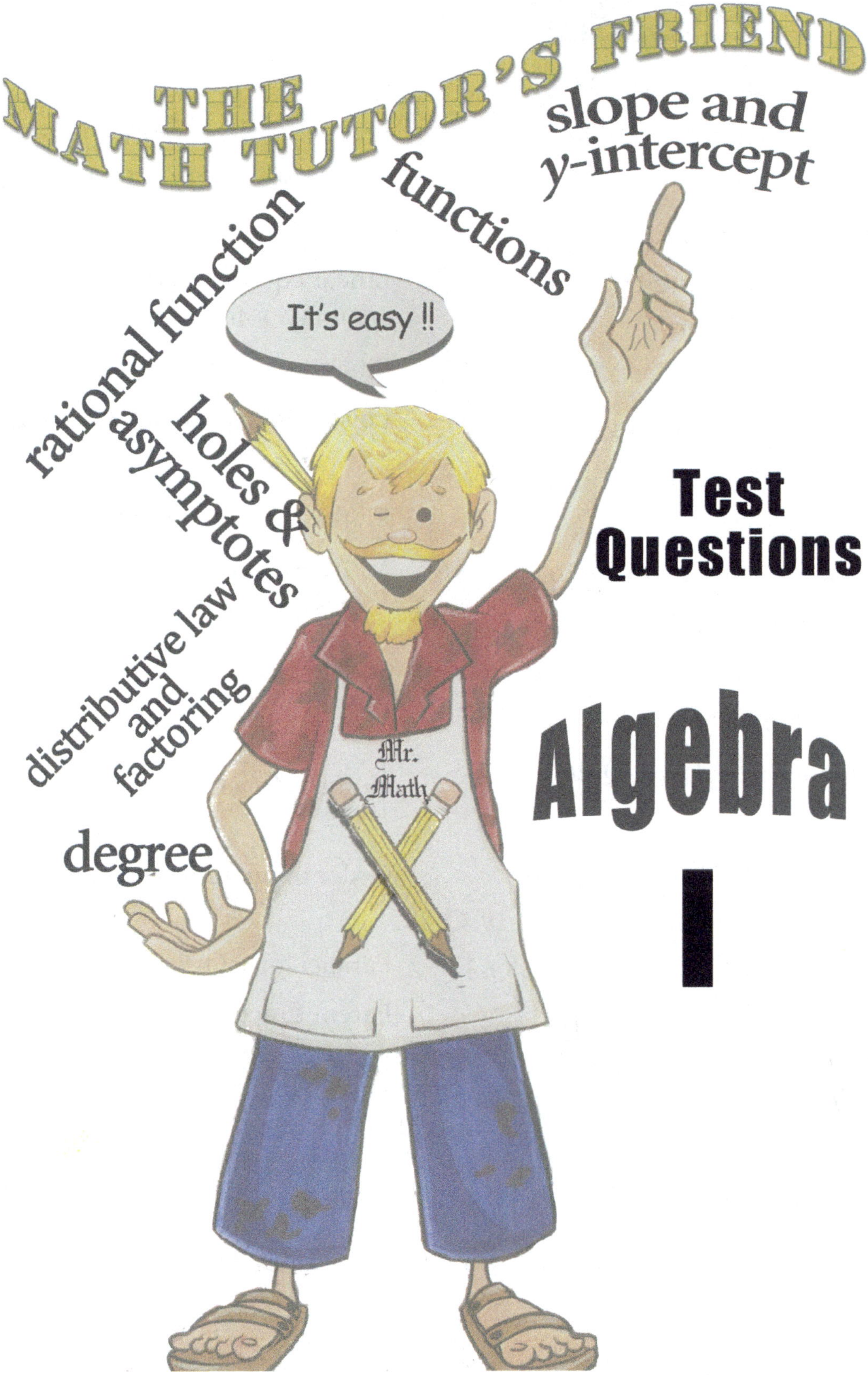
THE MATH TUTOR'S FRIEND
slope and y-intercept
functions
rational function
It's easy !!
holes & asymptotes
distributive law and factoring
degree
Mr. Math
Test Questions
Algebra I

QUIZZES

•

• The x-intercept of the line $y = \frac{1}{2} x + 1$ is

A. 1

B. ½

C. -2

D. 0

E. None of the above

• • Answer C. To find the x-intercept, set $y = 0$ (when the line crosses the x-axis, the value of y is zero) and solve for x:

$$0 = \tfrac{1}{2} x + 1;$$

add -1 to both sides to get $-1 = \frac{1}{2} x$; multiply both sides by 2 to get $-2 = 2 \times \frac{1}{2} x = x$, so that $x = -2$ when $y = 0$.

• ♥ •

• How many terms are there in the following equation ?

$$y - 11 = 8xa_{ij}^{4} - 2xa_{ij}^{2} + 8/[(x - 2)(x + 4)]$$

A. 2

B. 7

C. 4

D. 5

E. None of the above

• • Answer D.

There are five terms; we in include terms on both sides of the equation. Note that the ratio (or rational expression) $8/[(x - 2)(x + 4)]$ is a single term.

• Solve the inequality for y: $-2y + 3 \le 2x - 3$

A. $y \ge x$

B. $y \le -4x + 6$

C. $y \ge x - 6$

D. $y \ge -x + 3$

E. None of the above

• • Answer D. The steps are

a) add -3 to both sides: $-2y \le 2x - 6$

b) divide both sides by -2: $y \ge -x + 3$

(this changes the sense of the inequality)

• ♥ •

• The distance between the points (2, 8) and (5, 4) is

A. 3

B. 4

C. 5

D. 6

E. 25

• • Answer C.

This easily can be calculated mentally: the x-distance between the points is $5 - 2 = 3$; the y-distance is $4 - 8 = -4$ (distance 4 in the $-y$ direction). The squares of these are 9 and 16, respectively, and their sum is 25. The distance between the points is the square root of 25 which is 5.

• Convert 2,350% to a decimal and 35% to a fraction

A. 23.5 and $^{0.35}/_{100}$

B. 2,350 and $^{35}/_{100}$

C. 23.5 and $^{7}/_{20}$

D. 2.35 and $^{35}/_{100}$

E. None of the above

• • Answer C.

To convert a % to a decimal remove the "%" symbol and move the decimal point two places to the left. 2,350% becomes 23.5

To convert a % to a fraction remove the "%" symbol and divide by 100.

$$35\% = {}^{35}/_{100} = {}^{7}/_{20} \text{ (lowest terms)}$$

• ♥ •

• Assume a 7% sales tax. What is the price, P, of a taxable item sold for $100.00, tax included?

A. $100.07

B. $107.00

C. $93.00

D. $93.46

E. None of the above

• • Answer D.

$P + 0.07P = \$100$; $(1 + 0.07)P = \$100$; $1.07P = \$100$;

$P = \$100/1.07 = \$93.46$.

• The function $y = 3x^5 - 2x^2 + 10$ has

A. seven solutions

B. five solutions, at least one of which is real

C. two real solutions

D. three solutions, both of which are real

E. None of the above

• • Answer B. It is an equation of degree 5 so has five solutions. It is an odd function (the exponent with the highest degree is odd) so at least one of its solutions must be real. In fact, because complex solutions come in conjugate pairs, there are 1, 3, or 5 real solutions.

• ♥ •

• The line $y = 2x - 5$ is

A. perpendicular to the line $y = -2x - 5$

B. parallel to the line $-2x - 5$

C. perpendicular to the line $y = -½\,x$

D. parallel to the line $x - 5$

E. None of the above

• • Answer C:

$y = -½x$ is the equation of a line with slope $-½$ and y-intercept at the origin ($y = -½x + 0$). The line perpendicular to $y = ½x$ at the origin is $y = -2x$.

- The two points (2, 7) and (7, 2)

A. Have a separation of 9 units in the x-direction

B. Have a separation of 5 units along the y-axis

C. Have a separation of 2 units in the x-direction

D. Have a separation of 7 units along the y-axis

E. None of the above

Answer B.

They are a distance of 5 units in both the x and y directions.

• ♥ •

- The distance between the point (5, 8) from the horizontal line $y = 0$ is

A. 0

B. 3

C. 9

D. 8

E. None of the above

• • Answer D.

The y-value of the point is 8. The definition of the y-value of a point (x, y) is its distance from the x-axis.

- Two line segments whose end-points are equidistant are

A. at right angles to each other

B. rays

C. parallel to each other

D. obtuse

E. None of the above

• • Answer C.

• ♥ •

- The equation $(x - 4)/(2x + 1) + 2x^2 - 5x + (a - 2) = 0$ has

A. two expressions and a total of 5 terms

B. two expressions and a total of 9 terms.

C. one expression and a total of 6 terms.

D. one expression and 4 terms

E. None of the above

• • Answer A.

There are two expressions (one on either side of the " = " sign) and five terms four on the left (separated by operators) and one on the right. Note that "$(x - 4)/(2x + 1)$" is a single term.

• The line $y = x - 3$ is parallel to the line

A. $y = 3x + 1$

B. $x = y + 1$

C. $2x + y = 1$

D. $y = 2x + 2$

E. None of the above

• • Answer B.

In the slope-intercept form linear equation the slope is the coefficient of the variable x (after *like* terms, if any, are collected).

The equation in answer B converts to the slope-intercept form $y = x - 1$.

• ♥ •

• (1) Simplify $f(x) = (x^2 - 36)/(x + 6)$;

(2) what is the domain of $f(x)$?

A. (1) $f(x) = x - 6$; (2) $x \neq -6$

B. (1) $f(x) = x + 6$; (2) $x \neq -6$

C. (1) $f(x) = x - 6$; (2) all x

D. (1) $f(x) = x + 6$; (2) $x \neq 6$

E. None of the above

• • Answer A. This is a "pattern recognition" problem. The difference of two squares, say $a^2 - b^2$, always factors into the conjugate pairs $(a + b)(a - b)$. This also works backward: If you see the conjugate pairs $(a + b)(a - b)$ you can quickly simplify them to $a^2 - b^2$. It is very important to understand this because you run into it frequently in algebra, geometry, and trigonometry and other math areas. As to the problem, just identify the perfect squares and factor them into conjugate pairs:

$$f(x) = (x^2 - 36)/(x + 6) = (x + 6)(x - 6)/(x + 6) = x - 6, x \neq -6$$

The *domain* of $f(x)$ is "All x, $x \neq -6$" (because division by 0 is undefined, and thus not allowed); there is a "hole" in the graph of $f(x)$ at $x = -6$.

• The expression $(a^2 - 2)x^2 + 3ax + 4$ is a

A. trinomial

B. polynomial of four terms

C. polynomial of five terms

D. Polynomial of seven terms

E. None of the above

• • Answer A.

It is a trinomial which is a polynomial of three terms.

• ♥ •

• The expression equivalent to $7x + 3y + 4x - 4y$ is

A. $10xy$

B. $10x$

C. $11x - y$

D. $11x - 8y$

E. None of the above

• • Answer C.

Collect the like terms: $7x + 4x - 4y + 3y = 11x - y$

• Apply the Distributive Law to the expression $-3(2x - 3y)$

A. $(6x - 9y)$

B. $(2x - 3y)$

C. $(5x - 6y)$

D. $(-6x + 9y)$

E. None of the above

• • Answer D

$$-3(2x - 3y) = (-3)\cdot 2x - (-3)\cdot 3y = -6x + 9y$$

• ♥ •

• Simplify the expression $a^2x^3 - 2ax^2 + ax$ by factoring

A. $a(a^2x^3 - 2ax^2 + ax)$

B. $a^2x^2(a^2x^3 - 2ax^2 + ax)$

C. $x^2(a^2x^3 - 2ax^2 + ax)$

D. $a^2x(a^2x^3 - 2ax^2 + ax)$

E. None of the above

• • Answer E.

The factored expression is $ax(ax^2 - 2x + 1)$

• A line segment has end-point coordinates at (-5, 12) and (5, 2). The coordinates of the point that bisects the line segment (center of the line segment) are:

A. (0, 7)

B. (2.5, 6)

C. (-2.5, 6)

D. (7, 0)

E. None of the above

• • Answer A.

Using the mid-point formula, the x-value point is the average (mean) of the x-values of the end-points and the y-values the average (mean) of the end-point y-values. x: $(-5 + 5)/2 = 0$ and y: $(12 + 2)/2 = 7$.

• ♥ •

• Solve for x $(3x - 5)/4 = 2x + 5$

A. $x = 5$

B. $x = -5$

C. $x = 0$

D. $x = -4$

E. None of the above

• Answer B.

$3x - 5 = 8x + 20$	[Multiply both sides by 4]
$-5 = 5x + 20$	[Subtract $3x$ from both sides]
$5x = -25$	[Subtract 20 from both sides]
$x = -5$	[Divide both sides by 5]

• The slope of the line passing through the points (2, 0) and (0, -3) is

A. 1.5

B. 0

C. -1

D. -5

E. None of the above

• • Answer A. The slope is the change in y divided by the change in x:

$$[0 - (-3)]/(2 - 0) = 3/2 = 1.5$$

NOTE: The equation of this line can be determined by writing the slope-intercept equation: $y = 1.5x + b$. Since the line passes through (2, 0) we can write:

$0 = 1.5\times 2 + b$, or $0 = 3 + b$, from which $b = -3$ so the equation of the line passing through the given points is $y = 1.5x - 3$.

• ♥ •

• Factor the following expression completely

$$4x^2 - 9y^2$$

A. $4x + 9y$

B. $2x^2 - 3y^2$

C. $(x + y)(x - y)$

D. $(2x - 3y)(2x + 3y)$

E. Cannot factor the expression any further

• • Answer D.

This is the "difference of squares" pattern.

• A student simplifies $3(2x - 1) - (3x + 1)$ as follows:

Step 1: $3(2x - 1) - (3x + 1) = 6x - 3 - (3x + 1)$

Step 2: $6x - 3 - (3x - 1) = 6x - 3 - 3x + 1$

Step 3: $6x - 3 - 3x + 1 = 3x - 3 + 1$

Step 4: $3x - 3 + 1 = 3 - 2$ [Student's Answer]

A. The answer is correct.

B. The answer is incorrect; mistake in Step 1

C. The answer is incorrect; mistake in Step 2

D. The answer is incorrect; mistake in Step 3

E. The answer is incorrect; mistake in Step 4

• • Answer C. $6x - 3 - (3x + 1)$, not $6x - 3 - (3x - 1)$

• ♥ •

• Match each example with the relevant property of equality

A. Reflexive	(1) if a = b and b = c then a = c
B. Symmetric	(2) a = a
C. Transitive	(3) if a = b then b can replace "a" in any expression
D. Substitution	(4) If a = b then b = a
E. Multiplication	(5) If a = b then a + c = b + c
F. Addition	(6) If a = b then a × c = b × c

A___; B___; C___; D___; E___; F___

• • Answer: A2; B4; C1; D3; E6; F5

• The equation $ax^2 + bx + c = 0$ contains how many terms?

A. 2 terms

B. 3 terms

C. 4 terms

D. 6 terms

E. None of the above because it is an equation

• • Answer: C.

This is the familiar quadratic equation; it is a trinomial (three terms) and a term ("0") on the right side of the " = " sign.

• ♥ •

• The algebraic expression equivalent to

$$(5/2)\ (10x^3 - 2/5\ x)$$

is

A. $50x^2 - 2$

B. $10x^2 - 1$

C. $x\ (5x + 1)\ (5x - 1)$

D. $x^2(5x - 2)$

E. None of the above

• • Answer C.

Step 1: distribute 5/2 to get $(25x^3 - x)$

Step 2: factor out x to get $x(25x^2 - 1)$

Step 3: recognize the "difference-of-squares" pattern to get $x\ (5x + 1)\ (5x - 1)$

• The degrees of the four terms in the expression $y^2 - x^3/y + x^2 - 32$ are, respectively,

A. two, two, two, zero

B. two, two, two, two

C. two, three, two, negative two

D. two, one, two, two

E. None of the above

• • Answer A.

The degree of each term is 2; (3 – 1 = 2); 2; and 0 (the degree of a constant is always zero)

• ♥ •

Which is true for x in the inequality?

$$-4x + 3 \geq 2x - 6$$

A. $x \geq -3$

B. $x = 3$

C. $x \leq 3$

D. $x \geq 6$

E. None of the above

• • Answer E.

Solving for x yields $x \leq 3/2$

• Line A in the diagram is expressed as $y = x + 5$. Choose the equation that could be a reasonable equation for Line B.

A. $y = 0.6x + 2$

B. $y = -0.6x + 2$

C. $y = x$

D. $y = -x$

E. None of the above

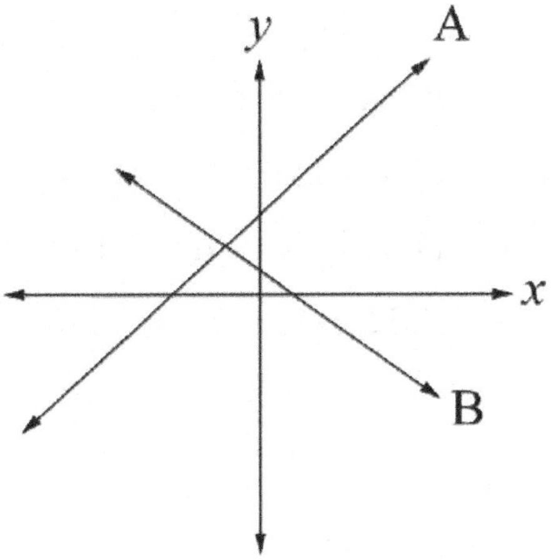

• • Answer B.

Line B has a negative slope and a positive y-intercept.

• ♥ •

• The diagram below shows that Y is a function of X.

A. TRUE - Y is a function of X

B. FALSE Y is not a function of X

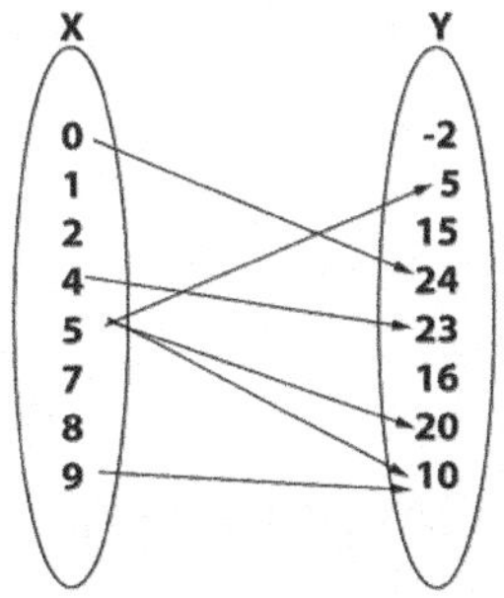

• • Answer FALSE.

$x = 5$ is related to more than one member of Y.

• An expression equivalent to $(x - 2)/y = 5/3x$ is

A. $3x(x - 2) = 5y$

B. $3x^2 - 6x = 5y$

C. $y = (3x^2 - 6x)/5$

D. All are equivalent

E. None are equivalent

• • Answer D.

Each is a step in solving for y.

• ♥ •

• The sum of two integers is 27. The difference is 5.

The numbers are

A. 27 and 5

B. 32 and 27

C. 16 and 11

D. 11 and 6

E. None of the above

• • Answer C. Let n and m be the two numbers. Then,

$m + n = 27$ [Given]

$m - n = 5$ [Given]

Add the equations from which $2m = 32$, and $m = 16$ and $n = m - 5 = 16 - 5 = 11$.

• If $x + 12 = 14$, evaluate $x^{3} - x^{-2} + 6x^{-1}$

A. 14

B. 12

C. 10.75

D. 11.25

E. None of the above

• • Answer C.

$$x = 2, \text{ so } x^{3} - x^{-2} + 6x^{-1} = 8 - \tfrac{1}{2}^{2} + 6/2$$
$$= 8 - 0.25 + 3 = 10.75$$

• ♥ •

• A theater sold tickets for a show: $15.00 for adults and $10.00 for "seniors." 150 tickets sold for $2,235. How many were "seniors" tickets? (Hint: Let A and S be the number of Adult and Seniors tickets sold, then S = 150 – A)

A. 75

B. 3

C. 15

D. No "senior" tickets sold

E. None of the above

• • Answer B. The number sold to senior citizens is 150 – A. Total prices for each group are 15A + 10(150 – A) = 2235

15A + 1500 – 10A + 15A = 2235	[Clear the parentheses]
5A = 2235 – 1500	[Collect like terms]
5A = 735	[Simplify]
A = 147	[Divide both sides by 5]
S = 150 – A = 150 – 147 = 3	[No. of Seniors tickets sold]

• The graph of a hyperbola shown below (dashed lines are not part of the graph)

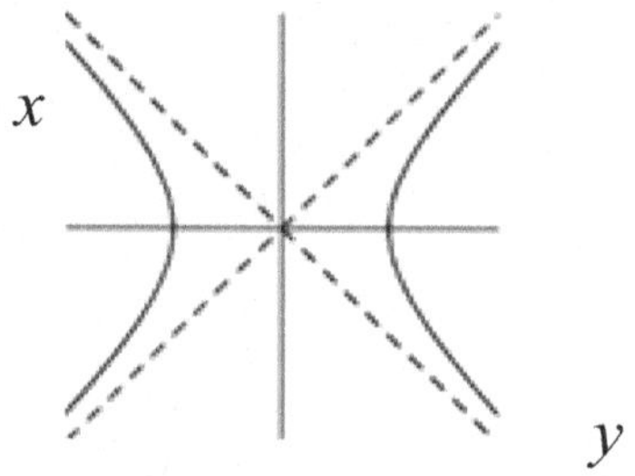

A. is a function

B. is not a function

C. may not be a function for certain values of x

D. has domain of all x

E. None of the above

• • Answer B.

There clearly are values of x that will have two associated values of y, as can be verified by the vertical line test. Therefore, the graph is not that of a function.

• ♥ •

• The parent function for the function $y = (x - 5)^3 + 13$ is

A. $y = (x - 5)^3$

B, $y = 13$

C. $y = x$

D. $y = x^3$

E. None of the above.

• • Answer D.

The function $y = (x - 5)^3 + 13$ is the parent function $y = x^3$ moved 5 units to the right in the x-direction and 13 units up in the y-direction.

• Let $y = \text{g}(x)$. Which is the independent variable?

A. x

B. y

C. g

D. All of the above

E. None of the above

• • Answer A.

The value of y depends upon the value of x selected from the domain of the function (i.e., the allowable value of x). Answer C is not correct because g is not a variable (it is a function).

• ♥ •

• Let $\text{f}(x) = x^2 + x - 2$. Then $\text{f}(3) =$

A. 10

B. -1

C. 5

D. 9

E. None of the above

• • Answer A.

If $\text{f}(x) = x^2 + x - 2$, Then f(3) means "substitute 3 for x in the equation."

$$\text{f}(3) = 32 + 3 - 2 = 9 + 3 - 2 = 12 - 2 = 10$$

• Let $f(x) = \sqrt{(x - 3)}$ (i.e., the square root of $(x - 3)$. The domain of this function is

A. All x such that $x > 0$

B. All x such that $x < 0$

C. The interval $[0, \infty)$

D. The interval $(-\infty, 3)$

E. None of the above

• Answer E. None of the above.

The domain of a function consists of all allowable values of x. Here, we have a square root function and the square root of negative numbers are not allowed. To find the domain, determine the values of x that will make the radicand $(x - 3)$ negative. Clearly, $x < 3$ will make the radicand negative, so the domain of $\sqrt{(x - 3)}$ is the interval $[3, \infty)$, that is, the domain is all x such that x is equal to or greater than 3.

• ♥ •

• Three consecutive odd numbers add to 57. What is the middle number?

• • Answer 19. Any odd number is an even number plus one: $2n + 1$;

the next two consecutive odd numbers are $2n + 3$ and $2n + 5$.

Thus, $2n + 1 + 2n + 3 + 2n + 5 = 57$

$$6n + 9 = 57; \qquad 6n = 48; \qquad n = 8$$

The middle odd number is $2n + 3 = 2\times8 + 3 = 19$.

(The three consecutive odd numbers are 17, 19, and 21 which add to 57)

• In math, the symbols " + " " – " " ÷ " and " × " are called

A. division

B. terms

C. operators

D. exponents

E. None of the above

• • Answer C.

The symbols represent the operations of addition, subtraction, division, and multiplication. Note: this can be reduced to the two operations of addition and multiplication because subtraction is adding a negative number and division is multiplying by a reciprocal. Thus, $4 - 7 = 4 + (-7)$ and $25 \div 5 = 25 \times 1/5$.

• ♥ •

• $[2(9 - 2) + 1] \div 3 + (2 \times 10) - 4 = ?$

A. 31

B. 20

C. 24

D. 21

E. None of the above

• • Answer D. Using the "Order of operations" rule:

$[2(9 - 2) + 1] \div 3 + (2 \times 10) - 4$

$= [2(9 - 2) + 1] \div 3 + 20 - 4$	[work inside parentheses]
$= (14 + 1) \div 3 + 20 - 4$	[clear inside parentheses]
$= 15/3 + 20 - 4$	[clear parentheses]
$= 5 + 20 - 4$	[do division]
$= 25 - 4 = 21$	[add & subtract left to right]

• 3 – 6 is to the right of 6 – 3 on the number line

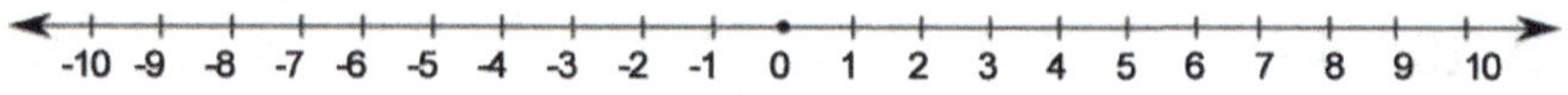

A. TRUE

B. FALSE

C. Cannot be determined with data given

• • Answer B.

3 – 6 = -3 is clearly to the left of 6 – 3 = 3 on the number line

• ♥ •

• On the graph shown below, points A and D are

A. real solutions to $x^3 - 3x^2 + 2$

B. zeroes

C. x-intercepts

D. All of the above

E. None of the above

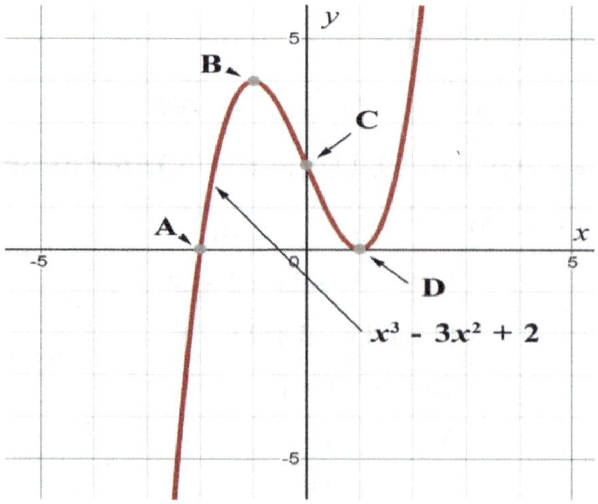

• • Answer D

• On the graph shown below, points B and D are

A. real solutions to $x^3 - 3x^2 + 2$

B. zeroes

C. local extrema

D. inflection points

E. None of the above

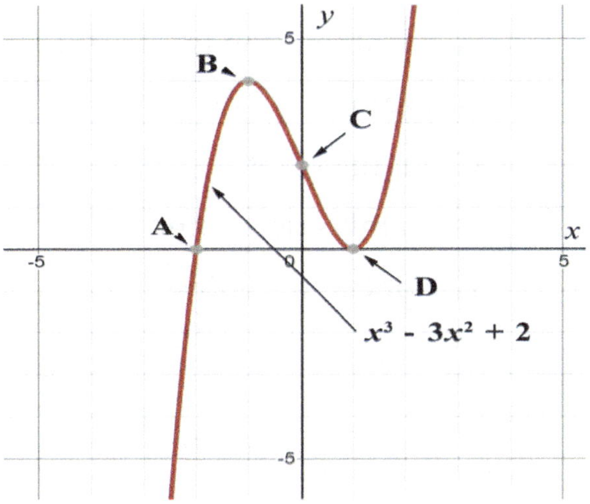

• • Answer C

• ♥ •

• On the graph shown above, the function has ___ solution(s)

A. 1

B. 2

C. 3

D. no solution

E. An infinity of solutions

• • Answer C.

The function has degree 3 therefore has three solutions: one solution (point D) has multiplicity two.

• On the graph shown below, point C is

A. a real solution to $x^3 - 3x^2 + 2$

B. a zero

C. a y-intercept

D. an inflection point

E. None of the above

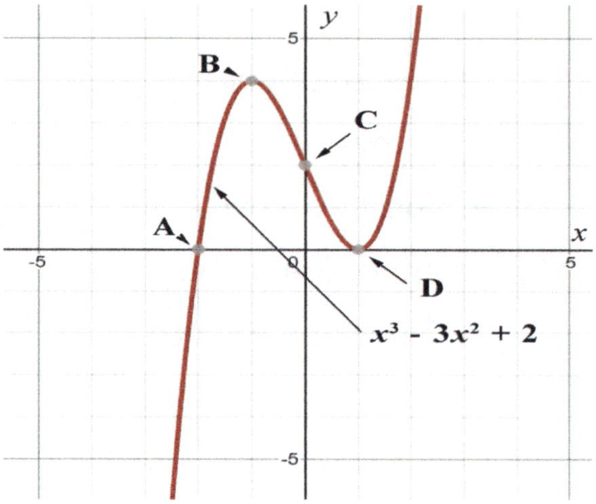

• • Answer C or D

It is both a y-intercept and an inflection point

• ♥ •

• The function $\dfrac{x(x-2)(x+1)}{x(x-1)(x+2)}$ is a _______ function.

A. rational

B. linear

C. quadratic

D. cubic

E. None of the above

• • Answer A.

The function is the ratio of two functions and thus is a rational function.

• The function .

$$\frac{x(x-2)(x+1)}{x(x-1)(x+2)}$$

A. has a horizontal asymptote at $x = -1$

B. has a "hole" at $x = -2$

C. has a "hole" at $x = 0$

D. has a vertical asymptote at $x = 2$

E. None of the above

• • Answer C.

The ratio x/x is undefined at $x = 0$

• ♥ •

• The function $\frac{x(x-2)(x+1)}{x(x-1)(x+2)}$

A. has a vertical asymptote at $x = -1$

B. has a vertical asymptote at $x = -2$

C. has a vertical asymptote at $x = 0$

D. has no asymptotes

E. None of the above

• • Answer B.

The denominator becomes zero at $x = -2$ (and also at $x = 1$) and the numerator does not.

The function has a "hole" at $x = 0$.

• The domain of the piecewise function shown is

A. $(-\infty, \infty)$

B. $(-\infty, 2]$ and $[2, 4]$ and $[4, \infty)$

C. $(-\infty, 2)$ and $(2, 4)$ and $(4, \infty)$

D. $(-\infty, 2)$ and $(2, 4)$ and $[4, \infty)$

E. $(-\infty, 2)$ and $(2, \infty)$

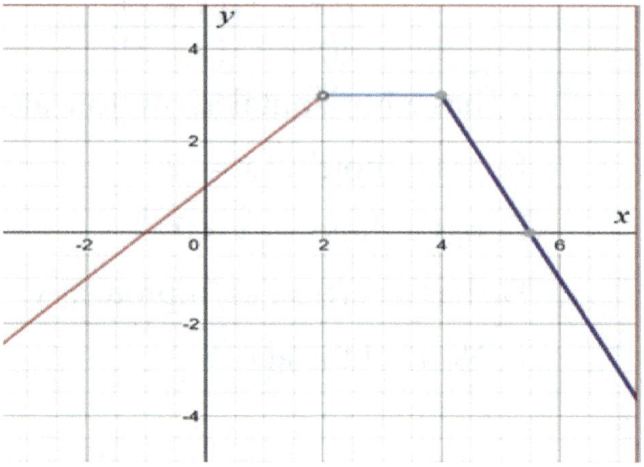

• • Answer E.

There is a hole at $x = 2$; all other x are in the domain.

• ♥ •

• The inverse, $f^{-1}(x)$ of the function $f(x) = (x + 2)$ is

A. $f^{-1}(x) = (x - 2)$

B. $f^{-1}(x) = (x - 1)$

C. $f^{-1}(x) = (x + 2)$

D. $f^{-1}(x) = (x - 1)/(x + 2)$

E. None of the above

• • Answer A.

Remember, $y = f(x)$. Just swap x and y and solve for y.

• The reciprocal, $[f(x)]^{-1}$ of the function $f(x) = (x + 2)/(x - 1)$ is

A. $(x + 2)/(x - 1)$

B. $(x - 1)$

C. $(x + 2)$

D. $(x - 1)/(x + 2)$

E. None of the above

• • Answer D.

$$[f(x)]^{-1} = 1/f(x) = (x - 1)/(x + 2).$$

• ♥ •

• Solve the systems of equations:

$$2y = 3x + 22$$

$$y = -5x - 2$$

A. (0, 0)

B. (-2, 8)

C. (2, -8)

D. (2, -2))

E. None of the above

• • Answer B. Multiply lower equation by -2, then add the equations. Solve for x and substitute the x-value in either equation to solve for y. Express the answer as the point of intersection at (x, y).

• Let $f(x) = 2x^3 - 4x^2 + 5x - 6$

$$f(3) = ?$$

A. 31

B. -16

C. -27

D. 27

E. None of the above

• • Answer D.

Replace x with 3 throughout and do the arithmetic.

• ♥ •

• Rationalize the function $x^2/\sqrt{(3x)}$

A. $3x^3$

B. $(x\sqrt{3x})/3$

C. $\sqrt{(3x^3)}$

D. $x^2\sqrt{(3x)}$

E. None of the above

• • Answer B.

Multiply by $\sqrt{(3x)}/\sqrt{(3x)}$ and simplify

• ♥ •

• The conjugate of $1 - x$ is

A. $1 - x$

B. $x - 1$

C. $1 - x^2$

D. $x^2 - 1$

E. $1 + x$

• • Answer E.

Just change the operator sign of the binomial

• $x^{-3} = 1/x^{3}$

A. True

B. False

• • Answer A.

A negative exponent signifies reciprocal

• ♥ •

• The amount that $100 will accumulate to after 3 years at 6% compound annual interest is:

A. 100.06

B. 106.00

C. 119.10

D. 118.06

E. None of the above

• • Answer C.

$$100 \times 1.06 \times 1.06 \times 1.06 = 100 \times (1.06)^{3} \approx 119.10$$

• Convert log(3) to base ***e*** to three decimal places

A. 0.500

B. 10.000

C. ***e*** (≈ 2.718...)

D. 0.477...

E. None of the above

• • Answer D.

$$\log(3) = \log_{10}(3) = \ln(3)/\ln(10) = 0.477...$$

• ♥ •

• Rationalize the denominator and simplify:

$$10/\sqrt{12}$$

A. $12/\sqrt{12}$

B. 5/6

C. $(5\sqrt{3})/3$

D. $(10\sqrt{12})/12$

E. None of the above

• • Answer C.

$$10/\sqrt{12} = 10/(2\sqrt{3}) = 5/\sqrt{3} = 5\sqrt{3}/(\sqrt{3}\times\sqrt{3}) = 5\sqrt{3}/3.$$

Note that answer D is correct, but not simplified. as required.

• $\ln(x) = a$ is equivalent to

A. $\log_{10}(x) = a$

B. $\boldsymbol{e}^{x} = a$

C. $\boldsymbol{e}^{a} = x$

D. $\boldsymbol{e} \times x \times a$

E. None of the above

• • Answer C.

Raise both sides to base ***e***.

$$\boldsymbol{e}^{\ln(x)} = \boldsymbol{e}^{a} = x$$

or apply the definition of logarithm

$$\boldsymbol{e}^{a} = x$$

• ♥ •

• $x^{(a + b + c)} =$

A. $x^{a} + x^{b} + x^{c}$

B. x^{abc}

C. $x^{a}x^{b}x^{c}$

D. All of the above

E. None of the above

• • Answer C.

This is one of the Laws of Exponents.

• $(xyz)^2 =$

A. $x^2y^2z^2$

B. $x^2 + y^2 + z^2$

C. $x^{-2}y^{-2}z^{-2}$

D. Any of the above

E. None of the above

• • Answer A.

• ♥ •

• Elementary row operations on an augmented matrix

A. have corresponding operations on the related system of equations

B. are not possible because the matrix is augmented

C. result in non-zero entries

D. involve squaring the augmented matrix

E. None of the above

• • Answer A.

• The function 7^x

A. is an exponential function

B. Has y-intercept at $y = 1$

C. Has a horizontal asymptote at $y = 0$

D. All of the above

E. None of the above

• • Answer D

• ♥ •

• The product of a square matrix **B** and its inverse is

A. $\mathbf{B}^{T}$ (transpose of **B**)

B. 1/**B**

C. ***I*** (Identity matrix)

D. All of the above

E. None of the above

• • Answer C.

$$\mathbf{B}\mathbf{B}^{-1} = \boldsymbol{I}$$

• The determinant of an identity matrix is

A. 0

B. 1

C. undefined

D. depends on the dimension of the matrix

E. None of the above

• • Answer B.

• ♥ •

• A singular matrix has determinant

A. 0

B. 1

C. undefined

D. depends on the dimension of the matrix

E. None of the above

• • Answer A.

• **A** is a 3×4 matrix and **B** is a 4×3 matrix. The product **AB** is

A. not possible because the matrices are not conformable for multiplication

B. 3×4

C. 3×3

D. 4×4

E. 4×3

• • Answer C.

The dimension of the product of an m×n matrix and an n×m matrix is m×m.

• ♥ •

• The determinant of the following matrix

$$\begin{matrix} 1 & 4 & 2 & 5 \\ 0 & 3 & 3 & 1 \\ 1 & 1 & 0 & 4 \end{matrix}$$

A. is 23

B. is 24

C. is 0 because the matrix is singular

D. does not exist

E. None of the above

• • Answer D.

Because the matrix is not a square matrix.

• The identity matrix is characterized by

A. a "1" in a row and zeroes elsewhere

B. at least one "1" in the main diagonal, and zeroes elsewhere

C. all values of "1" in at least one row

D. all values of "1" in at least one column

E. None of the above.

• • Answer E.

The identity matrix consists of zeroes except for "1"s in the main diagonal.

• ♥ •

• The end-behavior (as $x \to \infty$) of the rational function

$$y = (2x^3 - x^2 + 5x)/(3x^2 + 45x - 20)$$

is

A. $y = 0$

B. $y = 2/3$

C. $y = 7/68$

D. $y = +\infty$

E. None of the above

• • Answer D.

The degree of the numerator is larger than that of the denominator

• In a matrix **A**, the notation a_{43}

A. the dimension of the matrix (4×3)

B. is the element of **A** that is in the 4th row and 3rd column

C. is equal to 1

D. is the element of **A** that is in the 3rd row and 4th column

E. None of the above

• • Answer B.

• ♥ •

• Which of the following functions has a slant asymptote?

A. $(3x^2 + 5)/(x^3 - 2)$

B. $x^2/(x - 1)$

C. $x^2/(4x^2 + 2)$

D. All of the above

E. None of the above

• • Answer B.

A: The degree of the denominator is 1 less than that of the numerator.

$(3x^2 + 5)/(x^3 - 2)$ has a horizontal asymptote at $y = 0$

$x^2/(4x^2 + 2)$ has a horizontal asymptote at $y = ¼$

• The function x^3 is symmetric

A. True

B. False

• • Answer A. It is symmetric about the origin

• ♥ •

• An *extremum* of a function is

A. an asymptote

B. a “hole” in the function

C. a maximum or a minimum point

D. an inflection point

• • Answer C.

• A local minimum for a function

A. may also be a global maximum
B. is a minimum near the origin
C. is always followed or preceded by a local maximum
D. is an asymptote

• • Answer C.

Else it would be a global minimum

• ♥ •

• The parent function of $y = f(x - h) + k$ is

A. $y = f(x)$
B. $y = k$
C. $y = h$
D. $y = mx + b$

• • Answer A.

It is a translation of $f(x)$ by h units to the right and k units up.

• The system of equations show below

$$y = 3x + 1$$

$$y = 3x - 1$$

A. has no solution

B. has an infinite number of solutions

C. has one solution

D. has two solutions

E. None of the above

• • Answer A.

The slopes are equal so the lines do not intersect. If the operator sign in the second equation matched that of the first, then the system would have an infinite number of solutions.

• ♥ •

• A quadratic equation with real zeroes is

A. a linear equation

B. a special case of the cubic equation

C. the product of two linear factors

D. undefined

E. None of the above

• • Answer C.

• The graph of $ax^2 + bx - c$

A. has a shape that depends on the value of c

B. is always "U" shaped (may be upside down)

C. has two local extrema

D. always intersects the x-axis

E. None of the above

• • Answer B.

• ♥ •

• The value of y for the system of equations shown below is

$$y + 1.2x = 6$$

$$y = 200 + 1.2x$$

A. 1.2

B. 200

C. 7.2

D. 103

E. None of the above

• • Answer D.

Isolate y in the first equation then add the two equations to get $2y = 206$ from which y easily is determined to be 103.

• The slope-intercept form of the linear equation is

A. $y = \mathrm{m}x + \mathrm{b}$

B. $(y - y_1) = \mathrm{m}(x - x_1)$

C. $\mathrm{A}x + \mathrm{B}y + \mathrm{C} = 0$

D. Any of the above

E. None of the above

• • Answer A.

• ♥ •

• Divide $f(x) = 3x^4 - 2x^2 + 4x - 26$ by $(x - 3)$. The remainder is (hint: use synthetic division)

A. 21

B. 211

C. 0

D. -3

E. None of the above

• • Answer B.

Note that $f(3) = 211$ (the Remainder Theorem)

• The function $f(x) = (x - 1)^2$

A. is a parent function

B. is an odd function

C. is a translation of its axis of symmetry to the right

D. is a translation of its axis of symmetry to the left

E. None of the above

• • Answer C

• ♥ •

• The inverse of $f(x) = x^2$ is

A. $y = 1/ x^2$

B. $x = 1/ y^2$

C. $x^2 = y$

D. $y = \sqrt{x}$

E. None of the above

• • Answer D.

Remember, $y = f(x)$. Just exchange x and y then solve for y.

• An example of a vertical line is the equation

A. $y = a$

B. $y = x$

C. $x = y$

D. $x = a$

E. Vertical lines are undefined

• • Answer D.

The value of x is a for any value of y.

A is the equation of a horizontal line at $y = a$

B and C are diagonal lines of slope 1 passing through the origin

• ♥ •

• Parallel lines

A. have identical slopes

B. have identical y-intercepts

C. intersect at the origin

D. have slopes that are the negative reciprocal of each other

E. cannot be expressed by the slope-intercept form of the linear equation

• • Answer A.

• A line and its normal

A. have identical slopes

B. have identical y-intercepts

C. are parallel

D. have slopes that are the negative reciprocal of each other

E. cannot be expressed by the slope-intercept form of the linear equation

• • Answer D.

• ♥ •

• The point-slope form of the linear equation is

A. $y = \mathrm{m}x + \mathrm{b}$

B. $y = (x - 1)$

C. $y - y_1 = \mathrm{m}(x - x_1)$

D. All of the above

E. None of the above

• • Answer C.

From which we get the equation of a slope

• The reciprocal of the function f(x) is written as

A. f(x)$^{-1}$

B. f^{-1}(x)

C. f($1/x$)

D. All of the above

E. None of the above

• • Answer A.

• ♥ •

• The inverse of f(x) is written as

A. f(x)$^{-1}$

B. f^{-1}(x)

C. f($1/x$)

D. All of the above

E. None of the above

• • Answer B.

• Which is ***NOT*** correct?

A. $(f \circ g)(x) = f(g(x))$

B. $(f \circ g)(x) = g(f(x))$

C. $f \cdot g \neq f \circ g$

D. f and g are composite functions

• • Answer B.

• ♥ •

• The quadratic equation is

A. a second-degree equation

B. a slope-intercept equation

C. a fourth-degree equation

D. symmetric about the x-axis

E. None of the above

• • Answer A.

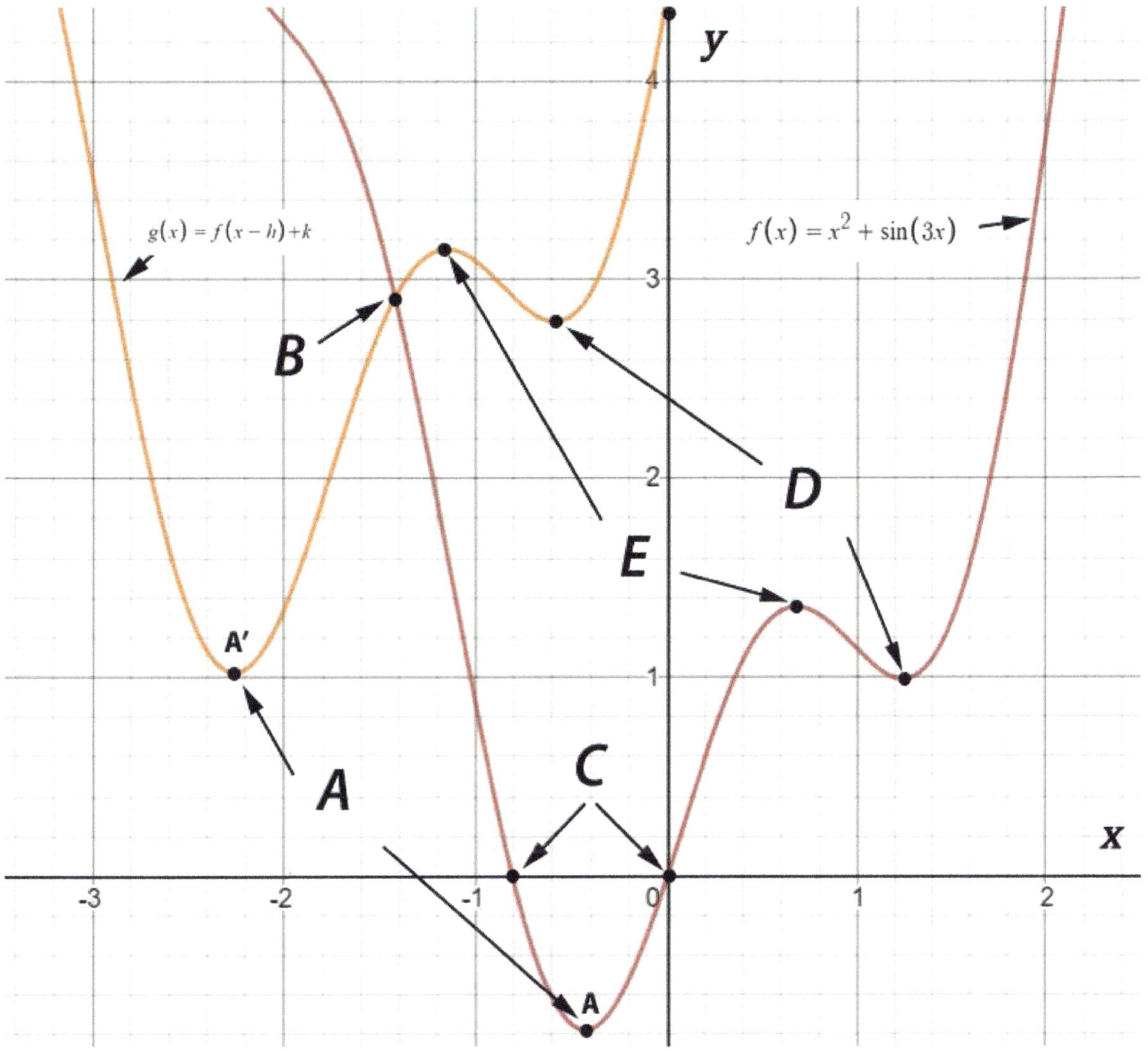

The following five questions refer to the graph above

[1] • Which set of points represent a global minimum?

A. A

B. B

C. C

D. D

E. E

• • Answer A

[2] • The point(s) where f(x) = g(x)

A. A

B. B

C. C

D. D

E. E

• • Answer B

• ♥ •

[3] • Which set of points represent real solutions to a function?

A. A

B. B

C. C

D. D

E. E

• • Answer C

[4] • Which set of points represent a local minimum?

A. A

B. B

C. C

D. D

E. E

• • Answer D

• ♥ •

[5] • Which set of points represent a local maximum?

A. A

B. B

C. C

D. D

E. E

• • Answer E

• $\ln(e) = ?$

A. 0

B. 1

C. e

D. π

E. None of the above

• • Answer B.

$\ln = \log_e$ and by definition of logarithm, $e^1 = e$

• ♥ •

• Write the system of equations represented by the graph below

$f(x) =$ ________________; $g(x) =$ ______________

A. x ; x^2

B. $x - 9$; x^2

C. $(x + 1)^2$; 9

D. $(x - 1)^2$; 9

E. None of the above

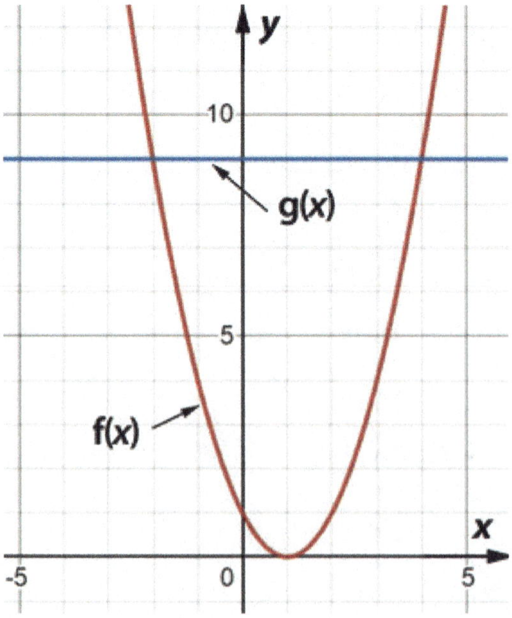

• • Answer: D.

$f(x) = (x - 1)^2$ and $g(x) = 9$

• A function has an axis of symmetry; this axis is

A. Always horizontal

B. Always vertical

C. May be horizontal or a line $x = y$

D. may be vertical or a line $y = x$

E. None of the above

• • Answer B.

If the axis of symmetry were not vertical (e.g., horizontal or diagonal) then the graph could not pass the vertical-line test and therefore could not be a function.

• ♥ •

• The expression $1 + 2 + 3 + 4 + 5 + \ldots + n$

is

A. a finite sequence

B. a partial sum

C. a finite series

D. an infinite sequence

E. an infinite series

• • Answer C.

The operator " + " identifies the expression as a series; the last term, n, is the last number of the finite series.

ANALYSIS OF THE FUNCTION

f(x) =

(label axes amd scales)

☐ Even function [f(x) = f($-x$)]

Axis of symmetry: x =

☐ Odd function [$-$f(x) = f($-x$)]

☐ Discontinuities: x =

☐ Holes:

☐ Vertical asymptotes:

☐ None

End Behavior

As x approaches $-\infty$ f(x) approaches ____________

As x approaches $+\infty$ f(x) approaches ____________

Parent Function (if any)

$y = x$	$y = x^2$
$y = x^3$	$y = \sqrt{x}$
$y = \lvert x \rvert$	$y = 1/x$
$y = a^x$	$y = \log(x)$
$y = \ln(x)$	$y = \sin(x)$
$y = \cos(x)$	$y = \tan(x)$

DOMAIN:

RANGE:

Increasing on

Decreasing on

☐ Global extrema

☐ Maximum:

☐ Minimum:

☐ Bounded above on y =

☐ Bounded below on y =

☐ Local extrema

Local maxima at (x,y):

Local minima at (x,y):

☐ Horizontal asymptote: y =

☐ Slant asymptote: y =

COMMENTS:

PARENT FUNCTIONS

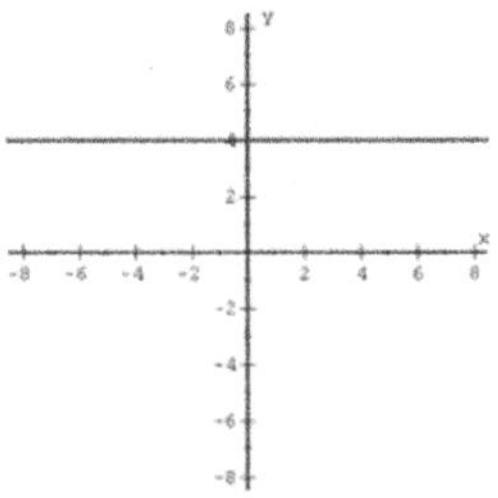

$f(x) = a$

Constant

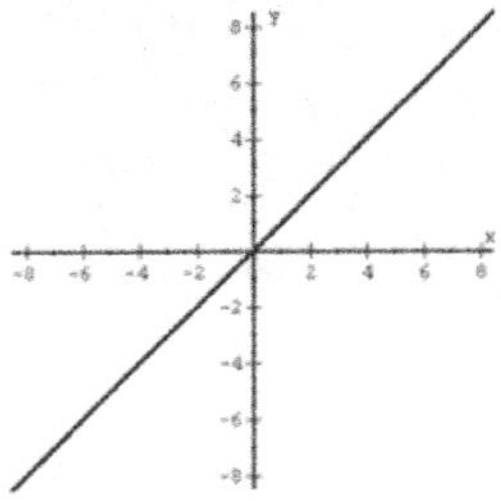

$f(x) = x$

Linear

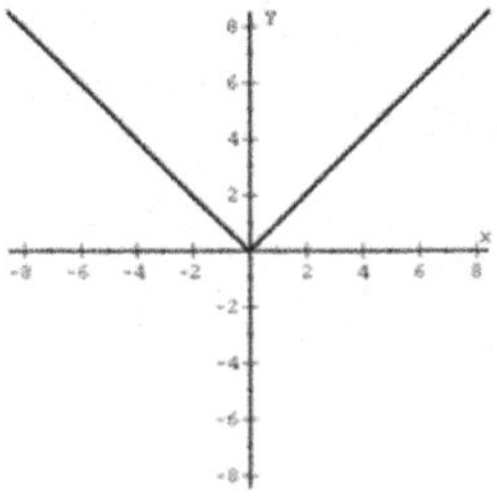

$f(x) = |x|$

Absolute Value

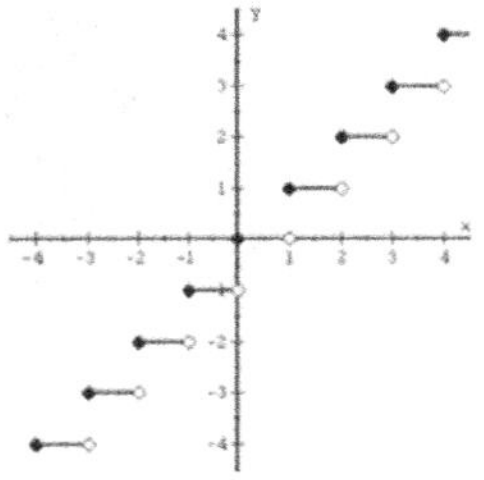

$f(x) = \text{int}(x) = [x]$

Greatest Integer

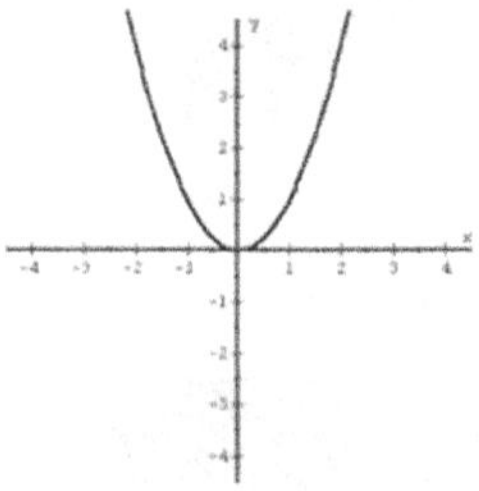

$f(x) = x^2$

Quadratic

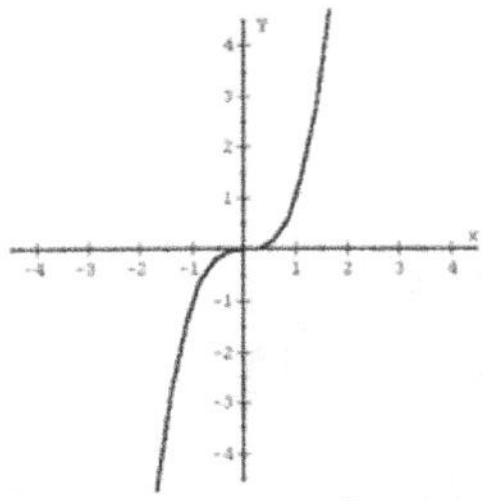

$f(x) = x^3$

Cubic

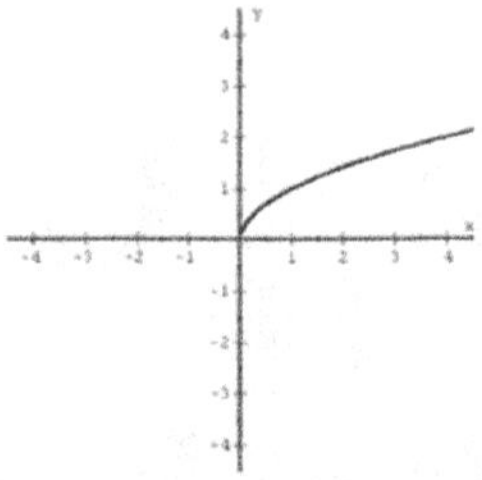

$f(x) = \sqrt{x}$

Square Root

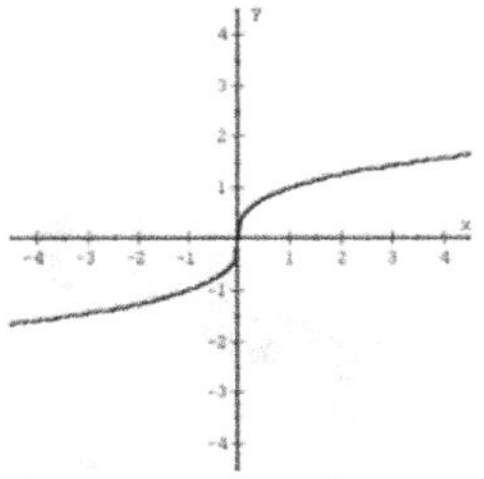

$f(x) = \sqrt[3]{x}$

Cube Root

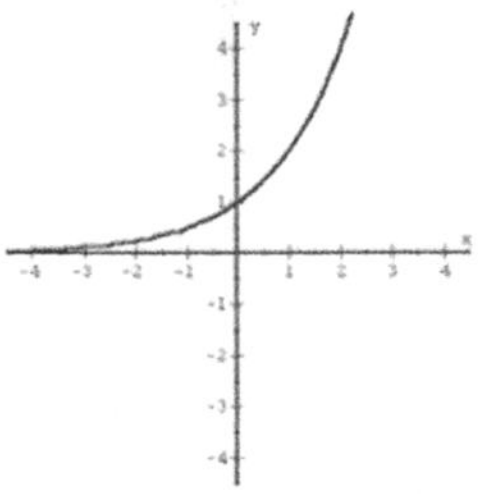

$f(x) = a^x$

Exponential

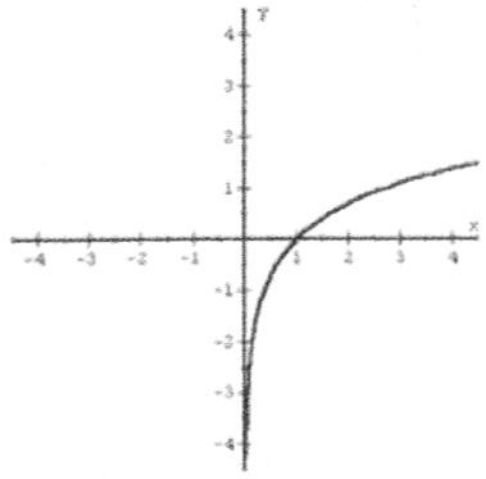

$f(x) = \log_a x$

Logarithmic

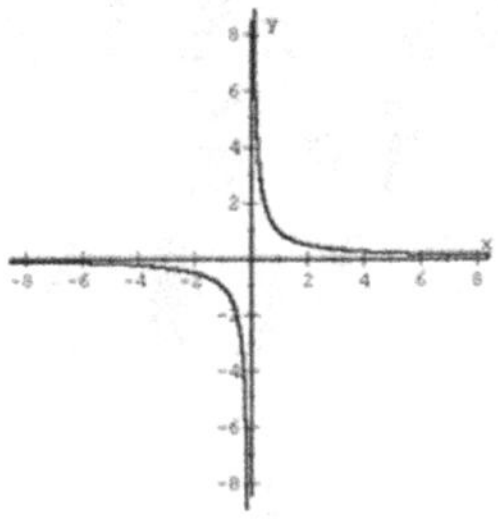

$f(x) = \frac{1}{x}$

Reciprocal

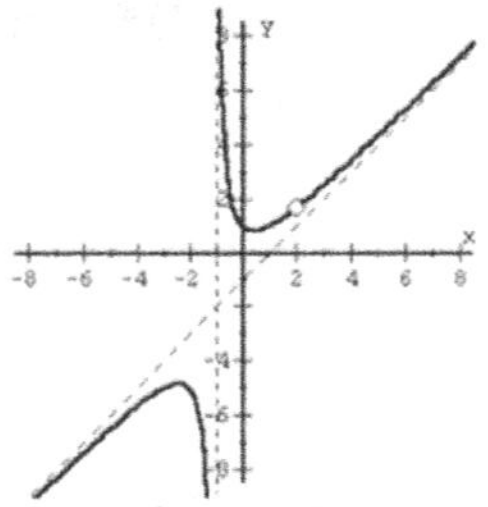

$$f(x) = \frac{(x^2+1)(x-2)}{(x+1)(x-2)}$$

Rational

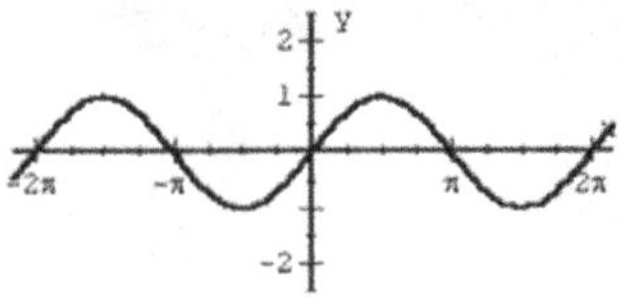

$f(x) = \sin x$

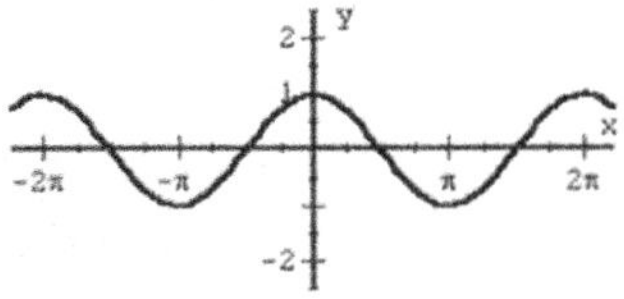

$f(x) = \cos x$

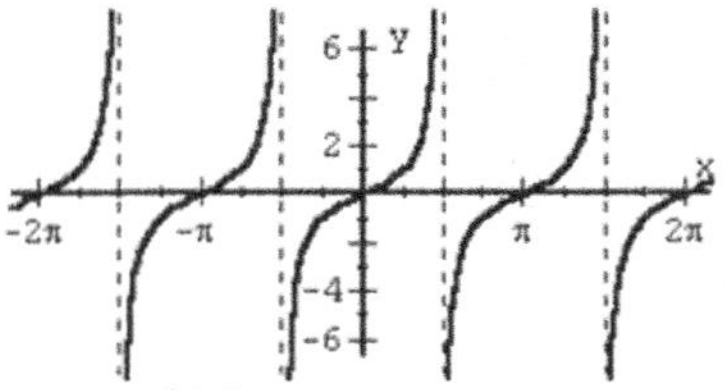

$f(x) = \tan x$

Trigonometric Functions

THE MATH TUTOR'S FRIEND

INFORMAL REVIEW OF HIGH-SCHOOL MATH TOPICS FOR TUTORS

-2-

• Algebra I •

Made in the USA
Monee, IL
02 November 2026

67381210R00096